Justina Chinossoca J. Chimuco

Processed Dairy Products

Justina Chinossoca J. Chimuco

Processed Dairy Products

Yogurt, cheese and butter production in dairy companies in
Luanda, Cuanza-Sul and Huíla Province (Angola)

ScienciaScripts

Imprint
Any brand names and product names mentioned in this book are subject to trademark, brand or patent protection and are trademarks or registered trademarks of their respective holders. The use of brand names, product names, common names, trade names, product descriptions etc. even without a particular marking in this work is in no way to be construed to mean that such names may be regarded as unrestricted in respect of trademark and brand protection legislation and could thus be used by anyone.

Cover image: www.ingimage.com

This book is a translation from the original published under ISBN 978-613-9-72769-8.

Publisher:
Sciencia Scripts
is a trademark of
Dodo Books Indian Ocean Ltd. and OmniScriptum S.R.L publishing group

120 High Road, East Finchley, London, N2 9ED, United Kingdom
Str. Armeneasca 28/1, office 1, Chisinau MD-2012, Republic of Moldova, Europe
Managing Directors: Ieva Konstantinova, Victoria Ursu
info@omniscriptum.com

Printed at: see last page
ISBN: 978-620-3-56475-4

CONTENTS

SUMMARY

In Angola, dairy products play a prominent role in people's eating habits. The aim of this study was to characterize the physical and chemical indicators of dairy products produced in Angola's industrial sector. It was carried out between February 2013 and October 2015. The process of making the products was monitored and the physicochemical parameters such as pH, acidity, energy, humidity, fat, protein and ash of yogurt, cheese and butter were determined. Company **A** was the largest producer of yogurt and butter, while Company **B** was the largest producer of fresh milk and cheese. The pH of the yogurt was 4.25 in Company **C**, 4.34 in B and 4.41 in A; acidity 80; 82 and 84 °D respectively and humidity 86.45; 86.37 and 89.40% respectively. The pH of the cheese was 4.72, 5.20 and 6.47; the energy value was 203.47, 403.92 and 416.39 Kcal; the moisture was 27.63, 42.80 and 44.57% and the fat 17.20, 38.30 and 39.68% respectively. The humidity of the butter from the three units was very high, varying between 37.90% for Empesa **C**, 46.87% for **B** and 47.30% for **C**, which indicates an excess of water. Protein was 1.00; 0.93 to 1.00%; ash 0.64; 0.73 to 0.99%. There were significant differences ($p \leq 0.05$) in the pH, energy, protein and fat values of the cheese and in the acidity, energy, fat and protein parameters of the natural yogurt from the three production units. As for butter, there were no significant differences between the protein and moisture values. The results indicate that not all dairy products produced in the industrial sector meet the parameters established by the international Codex Alimentarius standards in terms of physicochemical composition.

Key words: Yogurt, cheese, butter, physicochemical, quality.

I- INTRODUCTION

Milk is a natural product with high nutritional value, neither adulterated nor supplemented, which results from the regular, hygienic milking of healthy, well-fed female mammals (Lidon and Silvestre, 2007; Hartmann, 2009). Cow's milk is the most widely studied and produced milk in the world. However, other mammal species, such as goats, sheep, buffalo, camels, yaks and mare, among other animals, produce milk in sufficient quantity and quality to meet nutritional demands and the variety and attractiveness of dairy products, favoring milk consumption in different regions of the planet (Park *et al.*, 2007; Penna, 2011).

Milk is made up of water, proteins, sugars, mineral salts, fats and vitamins. Milk protein is suitable for the body to form and repair muscle tissue and is opaque white in color. Fat is a rich source of energy and gives milk its yellowish color (cream). The mineral salts found in milk, especially calcium, are important in the formation of bones and teeth, especially in children and pregnant women, contribute to various functions such as growth and nerve function, prevent osteoporosis and possibly colon cancer. Sugar is a source of energy (Ferreira, 2000; Ordóñez *et al.,* 2005; Bezerra, 2008).

Costa and Lobato (2009) state that milk is a unique food in terms of its nutritional value and composition. It is therefore an essential component of the diet of newborn babies for all mammal species, particularly humans, and is also recommended at all ages. For this reason, it is extremely important that it is presented with quality. Bezerra (2010) stated that most of the milk used to make dairy products is cow's milk, although milk from other mammals can also be consumed.

Milk intended for the manufacture of dairy products must also be of good quality, and this property is related to the herd's health conditions. Preservation and transportation are important stages in the processing of milk, where the quality levels of the products to be made are defined. All milk undergoes heat treatment and has different aspects depending on its purpose. These can be the production of yogurt, cheese or butter (Viana, 2012).

According to Lópes *et al.* (2012) and Viana (2012), obtaining good quality milk depends on the conditions in the dairy production units, as well as the care with which it is kept during storage. This quality has repercussions on the livestock, which contributes to the profitability of the dairy farm and, at the same time, the production of good, healthy milk that is rich in constituents.

According to Oliveira (2007), the factors that contribute to achieving these results include animal welfare, the process of obtaining milk and its conservation on farms before it is sent to the industries.

According to Brandão (2002) CEDGA (2009), the quality of the end product is intrinsically linked to the quality of the raw material. No matter how advanced the technology used in the process, it will never be possible to produce a good quality product from poor raw materials. Therefore, taking care to produce hygienic milk is of vital importance. Several factors must be observed in the industry before the products are processed. These include: facilities; personnel; utensils; conservation (refrigeration), animal health, cattle feed, mastitis, somatic cell count and the use of antibiotics.

Interest in healthy, nutritious and highly usable food products has grown worldwide, resulting in a number of studies in the field of dairy products. Some of these studies have emphasized the nutritional value of dairy ingredients, as well as the importance of a diet based on these products (Park *et al.*, 1997).

Dairy products have been known for millennia and it is quite likely that they have been used for human consumption since the time of the ancient nomadic tribes, due to the great availability of milk from the herds that were moved along with the human populations. The production of certain dairy products, such as cheese, is associated with popular culture and the culinary customs of cattle herders. Some authors mention that the same may have happened with the fermentation of milk that was stored in vessels made from the stomachs of animals (Reis and Combs, 2003).

Dairy products have historically developed in those human populations or communities that have evolved physically to maintain in adulthood a better ability to digest the main milk sugar: lactose (MADRP, 2007; Hartmann, 2009).

They are considered to be one of the main achievements of cultural evolution. Most of the lactose in milk disappears to be converted into other more digestible components after the lactic fermentation that takes place when it is made (Ferreira, 2001).

In Asia and Africa, dairy products are commonly consumed in only a few places, while they are more widely consumed in northern Europe and in areas of the world with a significant migratory presence from those regions, such as the United States of America and Australia (Bezerra, 2010).

The 20th century was the time when milk and dairy products saw a sharp rise in consumption across the planet. Advances in artificial milking methods, feeding and species selection, technological progress in transportation and refrigeration processes (Winck, 2010).

As already mentioned, the milk of various animals can be used as raw material to produce different types of dairy products on the market, such as yogurt, cheese and butter, which are considered to be the most widely accepted and consumed products (Bezerra, 2010). According to SENAR (2010), it is a food of excellent nutritional quality that can be used both *in natura* and processed into derivatives, the form of which adds value to the product, increases its useful life and diversifies the ways in which it can be consumed.

The dairy sector is characterized by its diversity of products and production lines. A dairy product is defined as a product obtained through any processing of milk that may contain food additives and functionally necessary ingredients. The dairy industry encompasses a large number of operations and activities that vary depending on the product. However, the fundamental operations are common to all production processes (Soroa, 2001; Winck, 2010).

According to Pupin (2002), the dairy industry is reacting to increase its competitiveness in the functional products segment, to adapt to the trend of changes in a demanding consumer market, which is changing rapidly, as well as having to maintain technological leadership in the food industry.

The physicochemical characteristics of dairy products are often tested in a similar way to milk, using lactometers to measure specific gravity. However, the preparation of dairy products differs according to the process that has been carried out; for example, some of them are subjected to lactic fermentation (such as yogurts), while others undergo a mechanical process to concentrate their fat content (butters). Sometimes a combined fermentation and maturation process is possible (cheeses). These processes change the composition and initial concentration of certain macronutrients and micronutrients, depending on the dairy product in question (Brandão, 1995; Abreu, 1997).

According to Faye and Konuspayeva (2012) more than 6 billion people in the world consume milk and dairy products; most of them live in developing countries. Since the early 1960s, *per capita* milk consumption in developing countries has doubled. However, milk consumption has increased more slowly than other animal products. Over the last two decades, *per capita* milk consumption has fallen in sub-Saharan Africa.

According to Sequeira *et al.* (2010), due to the inherent regional characteristics of dairy products, only 5% of world production is traded internationally. However, consumption is expected to increase in milk-deficient countries, and this trade is likely to expand in the coming years.

Data on the production and physico-chemical indicators of dairy products produced in the country is scarce, hence the importance of a survey to get a more realistic idea of the quality of the products produced in the areas under study.

1.1- Problem

Are the physicochemical indicators of dairy products (yogurt, cheese and butter) produced in Angola's dairy industry the most appropriate for obtaining acceptable quality?

1.2- Hypothesis

By evaluating the physicochemical indicators of dairy products (yogurt, cheese and butter) processed in the dairy industry, it will be possible to infer the quality and guarantee of the final product produced.

1.3- General Objective

Characterize the physicochemical indicators of dairy products (yogurt, cheese and butter) processed in the dairy industry.

1.4- Specific Objectives

1. Describe the situation of the dairy products processing industries (yogurt, cheese and butter);

2. To determine the physicochemical indicators of dairy products (yogurt, cheese and butter) processed in dairy industries;

3. To compare the physicochemical indicators of dairy products (yogurt, cheese and butter) processed in dairy industries;

II- LITERATURE REVIEW

2.1- Making dairy products

According to Gori (2012), in the food industry, food safety and quality are considered important issues worldwide and are directly related to health and social progress. Consumers are increasingly looking for food products from trusted brands and expect manufacturers to provide high quality products.

Milk is a balanced mixture of proteins, carbohydrates, fat, mineral salts and vitamins, dispersed in water as emulsions, colloidal suspensions and true solutions (Fonseca, 2000; Costa, 2011). It can be consumed as a liquid or transformed into various products such as cheeses, butter, yogurt, ice cream, creams and cream, among others (Teixeira and Fonseca, 2008; cited by Freitas and Maciel 2013).

Dairy products are prepared according to the following procedures: simple removal of water to produce totally evaporated or powdered milk, precipitation of certain components to produce certain varieties of cheese (which then undergo a maturing process in which there are considerable changes in lipids, carbohydrates and proteins), or chemical or biochemical modification of others such as yogurt and fermented milks (López *et al.,* 2012).

According to Ordóñez *et al.* (2005), the nutritional value of dairy products depends primarily on the milk, and consequently on the technological process, especially in terms of the physicochemical effects of nutrients, which deserve special attention due to the denaturing of sero-proteins and the destruction of vitamins.

Currently, heat treatments in many countries use temperatures of 135 to 150 ºC for 2 to 4 seconds, because they cause chemical changes for a sterilized effect. The denaturing of whey proteins affects milk coagulation, but not biological value or digestibility. Therefore, the interaction of carbohydrate proteins results in a decrease in lysine: another effect in relation to the treatments (López *et al.*, 2012).

2.1.1-Yogurt

The origin of yogurt can be traced back to the time when man began to domesticate species of dairy animals and use milk. Yogurt is one of the most popular types of acidified milk in the world. The advantages of replacing industrialized yogurt with homemade yogurt include the potential benefits of a functional food and a reduction in the cost of the product. The population is more aware of food and health, looking for the potential benefits of a functional food (Cidri *et al.*, 2005; Moleta, 2006).

Initially, yogurt consumption was quite limited, restricted to certain ethnic groups. In the mid-1960s, the addition of fruit to the product in order to soften its acidic taste led to greater acceptance, popularizing its consumption in all social classes and age groups. At the same time, its nutritional and therapeutic qualities were widely publicized, leading to a considerable increase in its consumption (Moreira *et al.*, 1999).

Yogurt is a nutritious fermented milk product that is being consumed in increasing quantities around the world. It originated with shepherds, particularly in Asia and southern and eastern Europe. Today, most researchers agree on the definition of yoghurt as coagulated milk, obtained through lactic acid fermentation between *Lactobacillus bulgaricus* and *Streptococcus thermophilus*, from pasteurized or concentrated milk, with or without the addition of powdered milk and with a consistency that differentiates it from other dairy products. It contains all the nutritional constituents of milk, with the exception of lactose, which is restricted during fermentation, making it advantageous for those who do not digest milk well (SENER, 2010; Martins *et al.*, 2013).

The Codex Alimentarius in Section 2.3 of the *General Standard for the Use of Dairy Terms* (CODEX STAN 206-1999) defines yogurt as a milk product, obtained by milk fermentation, which can be made from purchased milk products, with or without modification to its composition, and resulting in a reduction in pH, with or without coagulation.

Corazza (2006) states that industrialized yoghurt is also prepared using starter cultures, but its composition is modified because thickeners, colourings, preservatives and additives are added in the industry to increase its shelf life.

At the beginning of the 20th century, yogurt was considered a medicine and was only available in pharmacies. In recent years, its consumption has become widespread throughout the world, thanks to industrial, technological and, above all, scientific development. Several studies recognize the multiple nutritional virtues of yogurt and the presence of a series of multidimensional factors that interfere with the promotion of human health (Madeiro *et al.,* 2006).

Malta (2001) states that, in addition to the types of yogurt already considered traditional, such as flavored, liquid, with pieces of fruit and low-fat, the technological evolution of production has led to the entry of new concepts, with greater added value, which have progressively won over consumers. Due to their increasingly active and urban lifestyles, consumers have been including yogurt more frequently in their diets, not only because it is quick and convenient to consume, but also because of the organoleptic and health qualities associated with it.

As yogurt is a biological product with great dietary value, the addition of preservatives or fermentation inhibitors should be excluded in its handling, as they eliminate its therapeutic value, making it a product that is difficult to digest (Behmer, 1999; Corte, 2008).

The nutritional qualities of yogurt include its ability to re-establish the intestinal flora of the digestive system, providing better digestibility than milk, where enzymatic activity is associated with the growth of microorganisms that pre-digest the product. Yogurt can be eaten by lactase-deficient people (people who have difficulty digesting lactose), it is rich in lactic acid and reasonably in acetic acid, which are quickly absorbed by the body and play an active role in metabolism (Goulet, 1991; Rodas, 2001).

The bacteria that carry out fermentation are divided into two groups: homofermentative bacteria, which break down sugars, transforming them mainly into lactic acid which, due to the increase in acidity, coagulates the caseins in the milk; and heterofermentative bacteria, which in addition to lactic acid, also produce acetic acid, succinic acid, alcohols and gases. In this group of bacteria, pyruvic acid is also a metabolite resulting from the fermentation of these microorganisms, even leading to the formation of a curd depending on the time and temperature conditions, according to Tronco, 1997 (cited by Silva *et al.*, 2012).

2.1.1.1- Main raw materials in yogurt production

Although yogurt is better for digestion than milk, its main raw materials are milk and sugar.

2.1.1.1- Milk

It is the most complete food that nature provides. Chemically, it is defined as a complex mixture both in terms of the nature of its constituents and its physical state, with the following constituents: water 86.9%, protein 3.5%, fat 4.0%, lactose 4.9% and ash 0.7% (López *et al.*, 2012), as shown in Table 1.

Table 1- Average quantity of the main nutrients in common milks in 100 g

Milk (Origin)	Energy (Calories)	Proteins	Fat	Carbohydrates
Cow	63,0	3,1	3,5	5,0
Goat	92,0	3,9	6,2	5,4
Buffalo	115,0	5,2	8,7	4,3
Sheep	108,0	5,6	7,0	5,5

Source: Valsechi (2001).

2.1.1.1.2- Sugar

Sucrose is the most commonly used sugar in yogurt production. It can be added directly to the milk or, in the case of yogurt made with fruit pulp, added together with the pulp in the form of jam. If sugar is added to milk, it should be mixed with skimmed milk powder before heat treatment, as it is more soluble at high temperatures. In addition, heat treatment destroys the microorganisms present in the sugar, mainly yeasts, which can cause the packaging to swell during the product's shelf life (Madeiro *et al.*, 2006).

2.1.1.2- Industrial processing of yogurt

The first and perhaps most important stage in the production of yogurt is the selection of the raw material. Producers must make sure that the quality of the raw material is high and that it does not contain any impurities for production, i.e. the raw materials, particularly the milk, must have their characteristics controlled. It must contain a low content of bacteria, be free of enzymes and substances that slow down the development of the culture (e.g. enzyme inhibitors), as well as pathogenic microorganisms and antibiotics. After ensuring the existence of the respective raw materials, the process itself follows, which is subdivided into several phases (Baptista, 2003; Rodrigues, 2007; Robert, 2008):

- Mixing and homogenizing
- Pasteurization

- Fermentation

- Cooling

- Addition of fruit base

- Packaging

- Storage and preservation

2.1.1.2.1- Mixing and homogenization

The first stage of this process consists of adding the ingredients, such as sugar and fresh or powdered milk, to milk that has been previously analyzed for pH and fat content, in an airtight tank, i.e. a tank that does not allow any exchange of substances or microorganisms. The mixture is not always made with the same ingredients or in the same proportions, as it varies depending on the type of yogurt you want to make, low-fat, semi-skimmed or light (Baptista, 2003).

According to Rodrigues (2007), the preparation of whipped yogurt begins with the selection of milk, which must meet certain conditions such as:

- Acidity lower than 20 °Dornic;

- Normal aroma and taste;

- High soluble solids content;

- Absence of enzyme-inhibiting substances;

- Absence of pathogenic microorganisms;

- Standardized fat content.

The same author states that for yogurt to have a good consistency, the milk must have a defatted dry extract of 15%. To increase the solids content, powdered or concentrated milk can be added. It is at this stage that additives such as sweeteners and stabilizers (gelatine, agar-agar) are added to increase the sugar content and viscosity.

The main reasons for homogenization are to prevent the formation of solid agglomerates during incubation and to ensure that the milk fat is evenly distributed. Homogenization will lead to a reduction in the size of the fat globules, making the consistency smoother. To ensure that there is no excess water in relation to the other components, the mixture is heated to allow the water to evaporate. During the process, the milk drops in temperature from 90 to 70 °C (Silva *et al.,* 2010).

2.1.1.2.2- Pasteurization

After the mixing and homogenization phase comes pasteurization, whose main objective is to destroy pathogenic microorganisms and eliminate a large part of the milk's normal microbial flora, favouring the growth of the microorganisms that will be inoculated and the denaturing of whey proteins. This operation is usually carried out in heat exchangers at a temperature of 80 °C for 30 min or 90 to 95 °C for 5 min (López *et al.,* 2012).

2.1.1.2.3- Fermentation

At this stage, the culture of microorganisms that was prepared earlier in the laboratory is added to the milk. It is incubated for 4 to 5 hours at 42 to

43 °C for fermentation to take place. This process takes place in the vat itself and is maintained until the desired acidity is reached (Robert, 2008).

2.1.1.2.4- Cooling, filling and storage

After fermentation, in the case of whipped yogurt, the clot formed is broken up by stirring it in the vat, passing through a filter and then a plate heat exchanger, where it is cooled to a temperature of 4 or 6 °C. In the case of firm yoghurt, the fermentation phase takes place in the container itself so that a coagulum is formed which does not break. It is then packaged and stored in cold rooms, where the temperature is kept between 2 and 4 °C (Valsechi, 2001).

The processing of natural yogurt is similar to that of whipped yogurt, the main difference being in the fermentation stage. Before being incubated in greenhouses, the yogurt is packaged. The time taken to fill the individual packages should not exceed 30 minutes after inoculation to avoid problems with the consistency of the product. When it reaches between 90 and 95 °D, the product is stored in cold rooms at between 4 and 6 °C (Muhlbauer *et al.*, 2012).

2.1.1.3- Types of yogurt

According to Rodrigues (2007), depending on the physical-chemical nature of the clot, yogurt can be divided into: traditional, whipped and liquid. In traditional yoghurt, the fermentation process takes place inside the packaging itself, it does not undergo homogenization and the result is a product with a firm texture; in whipped yoghurt, the fermentation process takes place in fermenters or incubators with subsequent breakage of the clot, which promotes a creamy texture; and in liquid yoghurt, the mass is broken before cooling and the fermentation process takes place in tanks, generating fluidity in the final product.

There are several types of industrially manufactured yogurt (Ordóñez *et al.*, 2005):

➢ Firm consistency

➢ Natural

➢ Milkshake

➢ Liquid

➢ Colored and flavored

➢ With natural fruits and juices

➢ With pulp

➢ With cereals

➢ With cream, whole cream, partly skimmed and skimmed

2.1.2- Cheese

Cheese is a fresh or matured product obtained by the natural coagulation of milk or the action of specific enzymes, and separated from the whey. It is one of the oldest foods in history, dating back thousands of years BC (EMBRAPA, 2011).

It is a product made from milk and according to ancient records it has been used as food for over 4000 years. Until the middle of the 19th century, the cheese industry was built on farms where it was prepared from the surplus milk produced. The first cheese factory in America was built in New York in 1851 and since then, the cheese industry has undergone considerable evolution, particularly in terms of equipment (Mundim, 2008).

There are various stories about how cheese came to be, but it is not known for sure when it began to be made. It is believed that it was around 8,000 BC when they began to domesticate mammals in order to have a guaranteed source of milk or meat. It is not known whether it was in Europe, Central Asia or the Middle East (INMETRO, 2006; EMBRAPA, 2007).

In prehistoric times, before man could read or write, a legendary merchant traveler from Arabia, crossing a rustic mountainous section of Asia, already tired after a rough climb in the sun, took a break to restore his strength and feed himself. He had brought dried dates as food and a certain amount of goat's milk in a canteen made from dried sheep's stomach. But when he brought the canteen to his lips to sip the milk, only a thin, watery liquid oozed out of it. Curious, Kanana, the legendary traveler, cut open the canteen and saw, to his surprise, that the milk had turned into a white curd, not too unpleasant for a hungry man's palate. The rennet in the sheep's partially dried stomach had coagulated the milk, and the result of this chemical operation was cheese (Hohendorff, 2006).

According to EMBRAPA (2011), cheese rennet came about accidentally after milk was stored in containers made from stomachs or leather, where the enzymes contained in them coagulated the milk and a dense mass with a watery liquid was formed, which was removed to form a compact mass that could be eaten fresh or matured and stored for later consumption.

According to the Codex Alimentarius /FAO / WHO (2010) cheese is a solid or semi-solid product, fresh or matured, in which the value of the whey protein/casein ratio does not exceed that of milk and is obtained by partial or complete coagulation or by techniques involving coagulation of different types of milk or milk products, through the action of rennet, an enzyme (chymosin), in an acidic medium attributable to lactic acid or other coagulating agents and partially draining the whey resulting from this coagulation.

According to Prudêncio (2006), cheese is a dairy product produced in a wide variety of flavors and shapes all over the world. For Tamime (2006), the art of transforming milk into cheese is very old and is basically a process of concentrating milk in which part of the solid components, mainly protein and fat, are concentrated in the curd while the whey proteins, lactose and soluble solids are removed in the whey. The manufacturing yield and its centesimal composition are determined by the properties of the milk used, especially the composition and stages of the manufacturing process.

2.1.2.1- Cheese processing

In order to process the cheese, it is imperative to follow the steps below:

2.1.2.1.1- Raw material reception

At this stage, the milk is sanitized in order to eliminate impurities from the barn, and this operation is carried out in a stainless steel tank (Line *et al.*, 2005).

2.1.2.1.2- Homogenizing the milk

Homogenization of dairy milk is rarely carried out, only in the production of fresh cheese to increase cheese yield and in the production of Roquefort-type cheeses. By homogenizing the milk fat, the lipases of *Penicillium roquefortti* act more easily on the homogenized fat, lipolysing it more easily and thus developing the more intense flavours and odours characteristic of these cheeses (Valsechi, 2001).

2.1.2.2- Processing milk into cheese

Cheese making includes physical, chemical, biochemical and biological principles. Enzymology is very important, since it is enzymes that convert lactose into lactic acid and casein into curds, transforming proteins and sugars, the components responsible for the aroma, texture and flavor of cheese (Spreer, 1991).

Several stages are involved in converting milk into cheese, the main ones being: coagulation, acidification, desorption of the grain (syneresis), shaping and salting. By intervening in these stages, the cheesemaker can control the composition of the cheese, which will directly influence its maturation and the final quality of the product (Pernodet, 1990).

2.1.2.2.1- Milk coagulation

Milk coagulation can be enzymatic or acidic. Enzymatic coagulation of milk involves modification of the casein micelle by limited proteolysis (breaking of the Phe105 - Met106 peptide bond of κ-casein) caused by rennet enzymes or coagulants, followed by calcium-induced aggregation of these altered micelles. The curd formed has the appearance of a gel that occupies the same volume of milk used in the process. The rennet or coagulant is added to the milk normally at 32-35 °C in sufficient quantity for coagulation to take place in 30 to 40 min. The dose of rennet or coagulant varies according to the manufacturer and can be used in liquid or powder form, as long as it is diluted in non-chlorinated water and added slowly to the milk under agitation (Fox and Mcsweeney, 1998; Walstra *et al.,* 1999; Fox *et al.,* 2000).

According to Perry (2004); Neves-Souza and Silva (2005) the coagulation temperature depends on the starter culture and the rennet enzymes. The starter culture most commonly used in cheese making is mesophilus, which means it develops well at a temperature of 20 to 25 °C, while the rennet enzymes, responsible for coagulating the milk, work well at a temperature of 40 to 42 °C. The coagulation temperature should be regulated between 32 and 35 °C.

Rennet is mainly made up of two proteinases, chymosin, which is the enzyme of interest to the cheese industry due to its specificity for binding between amino acids 105-106 of κ-casein, and pepsin, which is a less specific, more proteolytic enzyme that is closely related to the bitter taste in cheeses (St-Gelais and Prudêncio, 2006).

Some factors that influence milk coagulation (Perry, 2004):

> Coagulation and gel formation do not occur at temperatures below 18 °C and above 60 °C, the rennet is inactivated;

> Temperatures close to 40 °C stimulate the action of the rennet and shorten the coagulation time;

> The closer to the optimum pH for the enzyme's action (approximately pH 6.0) the better the action of the rennet and the greater the strength of the curd;

> The greater the amount of soluble calcium present in the medium, the faster the clot will form and the firmer it will be;

> The higher the percentage of protein in the milk, the better it coagulates;

> The higher the concentration of enzymes, the shorter the milk coagulation time.

After coagulation, the curd network continues to form for a considerable time after a visible gel has been obtained, even after cutting. The strength of the gel formed is very important from the point of view of syneresis and, consequently, for moisture control and manufacturing yield (Fox and McSweeney, 1998).

Another coagulation technique is by acidifying the milk, in which the milk is left at room temperature and its acidity rises until it acquires a coagulated or "curdled milk" appearance. This system is used in the manufacture of various types of cheese (Walstra *et al.,* 1999).

2.1.2.2.2- Syneresis

The curd gel formed is quite stable, but shows syneresis when it is cut or broken. By controlling syneresis, the cheesemaker can easily control the moisture content of the cheese mass, the degree and extent of ripening and the stability of the cheese. The higher the moisture content of the cheese, the faster it will ripen, but the less stable it will be. Syneresis is promoted by the following factors (Fox and McSweeney, 1998 cited by Paula *et al.*, 2009):

- ➢ Lower cut thickness;
- ➢ Low pH;
- ➢ Presence of calcium ions;
- ➢ Increased cooking temperature;
- ➢ Stirring the curd during cooking;
- ➢ Higher protein and lower fat content.

2.1.2.2.3- Acidification

The primary function of the starter culture is the production of acid, which is crucial for the manufacture of most cheeses. The pH of the dough drops to around 5.0 in a time span of between 5 and 20 hours, depending on the variety of cheese to be made. Acidification is provided by the fermentation of lactose to lactic acid by lactic acid bacteria added to the milk or by direct acidification with the addition of lactic acid in some cases. Traditionally, cheesemakers relied on the endogenous microbiota present in raw milk for the fermentation of lactose, a process that is still used in the manufacture of Minas Artesanal cheese. However, this microbiota can vary greatly, changing the degree of acidification and, consequently, the quality of the cheese (Fox and McSweeney, 1998; Walstra *et al.,* 1999).

The use of whey is one of the most traditional ways of incorporating lactic acid bacteria into the manufacturing process and is widely used in the production of hard cheeses in Italy. The whey from the production of Parmesan cheese itself is removed after the pasta has been cooked at high temperatures (between 55 and 57 °C), leaving it to ferment until the following day where it will be used for the day's production. Its acidity can reach average values of 140 to 180 °D and is therefore a practical example of the natural thermal selection of lactic acid bacteria from the milk itself (Robinson, 2002).

Nowadays, freeze-dried yeasts are super-concentrated, highly active cultures for direct inoculation into the brewing tank. The use of this technology eliminates the preparation and handling of the yeast, thus eliminating problems and inconveniences caused by contamination. This type of yeast also allows for greater flexibility in production scheduling. Yeast guarantees a better and more uniform quality for the products produced, and is a modern technology now used in large industries and small dairies (Furtado, 2005).

There are various types of yeast on the market that can be used to produce different types of cheese. Those most commonly used in cheese making are usually composed of members of three genera (Robinson, 2002):

> ➢ The homofermentative mesophilic *Lactococcus lactis* subsp. *Lactis* and *Lactococcus lactis* subsp. *cremoris*, which are

acidifying bacteria suitable for cheeses with a low cooking temperature, with a closed mass and little aroma (Minas Curado, Prato for slicing, Mussarela);

> The heterofermentative mesophilic *Lactococcus lactis* subsp. *lactis* biovar *diacetylactis* and *Leuconostoc mesenteroides* subsp. *cremoris*, which are mesophilic flavoring bacteria that produce carbon dioxide and diacetyl (the main aroma compound in cheese and butter), are recommended for open-dough cheeses with a more pronounced aroma (Prato, Gouda, Cottage);

> The *thermophilic* ferments *Streptococcus salivarius* subsp. *thermophilus*, *Lactobacillus helveticus* and *Lactobacillus delbrueckii* subsp. *bulgaricus*, which are lactic acid bacteria from yogurt and are suitable for cooked or semi-cooked cheeses (Mozzarella, Provolone, Parmesan, etc.).

Acid production plays various roles in cheese making such as: it controls and prevents the growth of spoilage and pathogenic bacteria; it affects retention and coagulant activity during coagulation; it solubilizes calcium phosphate and therefore affects the texture of the cheese; it promotes syneresis and consequently influences the composition of the cheese and also the activity of enzymes during ripening (Ordóñez *et al.,* 2005).

2.1.2.2.4- Forming and pressing

When the final point of manufacture in the tank is reached, i.e. the desired pH and moisture content, the mass is separated from the whey and placed in molds of specific size and shape so that the whey drains between the grains and a continuous, homogeneous mass is formed. High-moisture cheeses form a continuous mass without needing to be pressed, but low-moisture cheeses need to be pressed with varying weights depending on the category (Line *et al.*, 2005).

Cheeses come in shapes and sizes ranging from 250 g to 80 kg. The size of the cheese is not only a question of aesthetics, but also a requirement for obtaining certain characteristics, with Emmental having to be large to avoid excessive diffusion of carbon dioxide gas, while Camembert must be small to allow mold enzymes to penetrate and thus provide more uniform ripening. In the case of pasta filata cheeses (Mussarella, Provolone, Cacio-cavalo), after fermentation, when the pasta has reached a suitable degree of demineralization (pH 4.8 to 5.2), it is heated in hot water, stretched and moulded into the desired shape. This process, called filleting, gives the cheese a very characteristic fibrous and filamentous structure (Fox and McSweeney, 1998).

2.1.2.2.5- Salting

Among the various stages of cheese making, salting stands out for its great importance, since salt (NaCl) has various functions in cheeses such as: flavor, control of microbial development, regulation of biochemical (enzymes) and physicochemical processes, durability, among others. Salting has a major influence on the final stage of production, which is ripening, since if it is not carried out properly, it can seriously affect the microbiological and enzymatic activity of a cheese and be the cause of various defects in the product (Fox *et al.*, 2000; Ordóñez *et al.*, 2005).

The most common salting methods are: in milk, in dough, in brine and dry. The average salt content of most cheeses ranges from 0.5 to 2.5%. In some cases, such as Creole cheeses, these values can be as high as 5 to 8% (approximately 15% salt dissolved in moisture). Regardless of the type of salting used, the salt used must always be of good physico-chemical and microbiological quality (Fox and McSweeney, 1998; Walstra *et al.,* 1999; Fox *et al.*, 2000; Costa *et al.*, 2004, cited by Paula *et al.*, 2009).

In addition to the functions of salting mentioned above, we can also list others, as reported by ILCT (1997); Viana (2012):

> ➤ It improves and enhances the taste, as well as masking foreign flavors;

> ➤ It attenuates the lactic taste of fresh curd and masks the pronounced lipolysis in moldy cheeses;

> ➤ Helps to form the rind of the cheese by surface dehydration;

> ➤ It helps to complement the desorption of the cheese, favoring the release of free water from the mass;

> ➤ It promotes the modification of osmotic pressure, the syneresis of the mass, stimulating the expulsion of whey and the reduction of moisture in the cheese.

According to Perry (2004), salting helps to control microbial growth and activity, providing a selection of the cheese's microbiota. Propionic bacteria do not tolerate low water activity, so Swiss cheeses are not left in brine for long. In the case of blue cheeses such as Gorgonzola, their higher salt content encourages the growth of mold.

2.1.2.2.6- Maturation

Maturing is the last stage of production and can last from a few hours to several months, depending on the type of cheese you want to make. During this process, a large number of aromas and flavors develop. It takes place in specially conditioned areas, where the temperature and humidity are suitable for each type of cheese, which means adapting climate-controlled facilities, although there are some regions that have natural temperature and humidity conditions that give their cheeses a special origin (Katiki, 2004 and Viana, 2012).

According to Tamime (2006), the particular characteristics of each type of cheese are defined during ripening. In addition, other factors also contribute to this determination, such as the macrobiota present, mainly from the yeast, milk and/or secondary cultures. The development of flavor and texture is strongly dependent on the pH profile, the composition of the milk and cheese, salting, ripening temperature and humidity.

According to Rocha (2004), during ripening, different varieties of cheese acquire their own flavor, aroma and texture characteristics through complex physical and chemical changes. During the process, the milk's own enzymes and those from rennet and microorganisms catalyze the decomposition reactions of the three main milk components: lactose, fat and protein, which are retained in the cheese. This decomposition results in numerous metabolites responsible for the variation in cheese characteristics.

2.1.2.2.7- Packaging and Labeling

Each type of cheese has its own packaging, which must guarantee its preservation and at the same time present the product well to the customer. There are various types of packaging, such as polyvinyl acetate-based plastic packaging and waterproof plastic in which the cheese is vacuum-sealed. The label must contain information about the product as well as the producer. The omission of information on the product is considered misleading advertising or a crime under the Consumer Protection Code (Gava, 2008).

2.1.2.3-Types of cheese

Depending on the time and conditions of ripening, cheese comes in three varieties (Bezerra, 2008):

- ➢ Frescal: unripened

- ➢ Half-ripe: 20 to 30 days of ripeness

- ➢ Cured: matured for at least 60 days

Unripened or fresh cheeses are those that are ready for consumption once they have been manufactured, therefore they contain high humidity and have an acidic or slightly sour taste. The best known are the chalet type, the so-called fresh cheese and cream cheese. Half-ripened cheese is neither fresh nor ripened, but matures for 30 days, acquiring a soft and firm consistency like Minas and Prato cheese (Katiki, 2004; Cardoso, 2006).

Cheeses that have been aged in different ways, and mainly evaluated for the growth of molds or for maturing bacterial cultures, are called aged cheeses. The enzymes produced by the growth of these microorganisms used in the curd produce the characteristic flavor and texture of these cheeses. The best known of these are Camembert and Roquefort, which can only be legally called by this name when it is produced in a French factory, those produced elsewhere are called blue cheese, because in their manufacture *Penicillium roquefortti* mold spores are added which is characterized by presenting a blue-green color through the cheese which gives it a bitter and pungent taste (Line *et al.*, 2005).

2.1.2.4- Grading cheese

The classification of cheese is often simplified by dividing it into:

- ➢ Soft cheeses
- ➢ Semi-smooth
- ➢ Firm.

As for ripening, the cheese is known as bacterial Munster. All these cheeses have a soft, smooth texture, high water or whey content and small rind formations. They require air when handled and are more difficult to keep than firm cheeses (Garcia, 2007).

The cheeses with a firm, ripe consistency are originally from Holland, obtained by enzymatic coagulation and very safe mechanical work, so those considered semi-hard are Gouda and Edam. The cooked cheeses, usually with holes, are slower to mature and are very easily preserved inside, such as Emmental and Gruyere (Lidon and Silvestre, 2007).

Cheeses can also be classified according to their fat content, humidity, texture and method of manufacture. An exact classification is limited since the same type of cheese can fall under more than one definition. Generally speaking, the main ones are:

1. According to fat content (%) (Ordóñez *et al.*, 2005):

> Fat

> Semi-fat

> Thin

> Skimmed

2. according to moisture content and texture (%) (MAPA, 2009):

> Low humidity cheeses (hard cheeses): humidity up to 35.9%;

> Medium-humidity cheeses (semi-hard cheeses): humidity between 36.0 and 45.9%.

> High-moisture cheeses (soft cheeses): moisture between 46.0 and 54.9%.

> Very high humidity cheeses (soft cheeses): humidity of not less than 55.0%.

3. According to cooking method and preparation (%) (Goetze, 2010):

> Minas Frescal, Minas Half Cured, Gorgonzola and Camembert

> Dish, Coberg, Gouda, Cheddar

> Parmesan, Swiss, Romano and Gruyere

> Mozzarella and Provolone

> Cottage cheese

> Ricotta

2.1.3- Butter

Butter is a fatty product in which the aqueous phase is dispersed in the oily phase forming a water/oil emulsion. It is formed by churning the cream previously obtained from skimming the milk. Among the components of milk, fat is the main element that goes into its manufacture (Costa, 2011).

According to CODEX STAN Standard 279-1971 (2011), butter is defined as a fatty product derived exclusively from milk, and/or from products obtained from milk mainly in the form of a water-in-oil emulsion.

According to Ministry of Agriculture, Livestock and Supply Ordinance (1996) No. 146 of March 7, 1996, butter is the fatty product obtained solely by churning and malaxing, with or without biological modification of pasteurized cream derived exclusively from cow's milk, using technologically appropriate processes. The fat content of butter must consist exclusively of milk fat.

Butter can be made from sour cream or not and may or may not contain salt. The oldest way of making butter was to let the cream acidify naturally and then beat it by hand. However, the natural acidification process is not very controllable and is very susceptible to contamination, which affects the quality and safety of the final product. Nowadays, the industrial production of butter is the result of the knowledge and experience that has been acquired in the areas of hygiene, bacterial physiology and heat treatment, as well as rapid technological development (Carvalho, 2008).

According to Valsechi (2001) and Fernandes *et al.* (2012), the main constituents of salted butter are: fat (80 - 82%), moisture (15.6 to 17.6%), salt (1.2%), protein, calcium and phosphorus (1.2%). Butter also contains fat-soluble vitamins (A, D and E).

Basically, the process of making butter involves concentrating the fat in the cream using a churn and removing the non-fatty phase (skimmed milk), which we call buttermilk (butter whey) (Augusta, 1998; Salinas, 2002).

The fat is separated by taking advantage of the difference in density between the fat and the other components of the milk. This operation is called skimming and is carried out using centrifuges. Among the centrifuges, the ones with the best results are hermetic, in which the percentage of fat in skimmed milk can be less than 0.04%, depending on the working temperature. The higher the temperature, the better the fat separation, but it must be borne in mind that if the milk has a high acidity, there will be partial coagulation of the casein and consequent clogging of the centrifuge. Temperatures between 30 and 35° C are often used. The quality of the cream depends on the composition of the initial milk (Valsechi, 2001).

2.1.3.1- Industrial butter-making process

According to the Portuguese standard 1711 and Ordóñez *et al.* (2005), the industrial production of butter involves the following stages:

- ➢ Cream reception;
- ➢ Neutralization;
- ➢ Pasteurization, deodorization;
- ➢ Inoculation with selected cultures;
- ➢ Maturation;
- ➢ Churning;
- ➢ Separation of the whey or buttermilk;
- ➢ Washing;
- ➢ Salting and kneading;
- ➢ Packaging and storage.

2.1.3.1.1- Neutralization or deacidification

This operation is carried out when the cream is highly acidic. Cream that is too acidic is thick and coagulates when pasteurized, leading to a burnt taste, as well as casein containing fat particles and many microorganisms during precipitation. This results in a drop in yield and a loss of efficiency in heat treatment (Ordóñez *et al.*, 2005).

According to Salinas (2002), under these conditions, the culture will develop abnormally, resulting in an oily butter with no consistency and bitter flavors. It is therefore necessary to reduce the acidity of the cream below 20° D for pasteurization to be processed normally. However, if deacidification is not carried out properly or if it is used excessively, saponification of the fat can occur, as well as creating conditions conducive to the proliferation of alkalizing or proteolytic bacteria, leading to unpleasant flavours.

Ordóñez *et al.* (2005) state that the most commonly used neutralizers are: sodium hydroxide, calcium oxide, calcium hydroxide, magnesium oxide, magnesium hydroxide, sodium carbonate, sodium bicarbonate, or a mixture of them. Calcium and magnesium salts don't cause foam and don't have a taste, but they are slow-reacting, have low solubility and produce viscous cream. Sodium hydroxide can cause saponification. Whatever neutralizer is used, it must have certain properties: chemically pure, finely powdered, previously diluted in water, added under constant stirring at the right temperature.

2.1.3.1.2- Pasteurization

The purpose of pasteurization is to prevent the spread of infectious diseases, reduce contamination in the industry and prevent possible alterations in the butter. It is carried out in the absence of air by keeping the cream at 92 to 95° C for 30 seconds. Under these conditions, the microorganisms and enzymes present are destroyed without altering the organoleptic qualities of the cream (Carvalho, 2008).

2.1.3.1.3- Deodorization

It is necessary to remove undesirable volatile substances from the cream that give the butter strange flavors and aromas. This treatment is called deodorization. There is a great deal of equipment designed to stimulate the deodorization of cream in the absence of air. This equipment, located at the exit of the pasteurizer, is under vacuum and receives the hot fat, which falls as a thin layer down its walls, causing the volatiles to escape (Valsechi, 2001).

According to Ralph (1998), after deodorization, the cream is sent to the refrigerator where it is rapidly cooled to the ripening temperature. The intensity and speed of cooling determine the size of the glyceride crystals with the highest melting point. If cooling is rapid, a large number of small crystals are formed, otherwise a smaller number of large crystals are formed, which increases the consistency of the butter.

2.1.3.1.4- Inoculation with *selected* culture

The heat treatment eliminates the natural flora of the cream, which must be replaced by a culture of selected lactic acid bacteria. Normally, the types of yeast used in butter production are: *Streptococcus lactis, S. cremoris, S. paracitrovorus, S. citrovorus* and *S. diacetylactis. S. lactis* and *S. cremoris* allow the acidity of the cream to increase, causing the precipitation of casein and providing conditions for the other slower-growing microorganisms responsible for the flavor and aroma of the butter (Clemente and Abreu, 2006).

The percentage of yeast added, under constant agitation, immediately after the cream has been refrigerated, varies from 2 to 5%, depending on the composition, the temperature of the cream and the maturation time. The acidity of the yeast is 85 to 90° D (Berticelli and Motta, 2011).

2.1.3.1.5- Maturation

Maturing causes a change in the structure of the cream which makes it easier to whip, as well as allowing the yeast to act. Ripening conditions depend on the composition of the raw material, the season and the desired characteristics of the final product. If the butter to be produced is for immediate consumption, the final acidity of the cream should be 45 to 58° D, which provides very pronounced organoleptic properties. When the butter is to be stored, the cream should have an acidity of 30 to 35° D, which gives the butter a discreet aroma and flavor, but provides a product of better quality and durability (Carvalho, 2008).

The same author states that in order to achieve the desired final acidity, in addition to the raw material, the percentage of inoculum added and the ripening time and temperature must be taken into account. At a temperature of 14 to 16° C, the ripening time is approximately 15 hours; under these conditions, aromatization is favored.

When the cream is rich in high-melting glyceride, small crystals need to be formed to make the butter softer, which is achieved by intense refrigeration at approximately 5° C for 4 hours. The cream must be heated from 13 to 15° C for the milk culture to develop. Maturation takes place in special double-walled tanks with agitators. There are units with a plate heat exchanger attached to the maturing tank. The cream circulates through the exchanger and returns to the tank until the desired temperature is reached, according to Salinas, 2002 (cited by Berticelli and Motta, 2011).

2.1.3.1.6- Churning

It is at this stage of production that the butter is formed. The cream, already prepared, is placed in the mixer, where the fat globules collide with each other and with the walls. As a result of these shocks, at a given temperature, the fat globules fuse together to form larger and larger agglomerates, breaking the balance of surface tension forces between the cream's components. As a result, the discontinuous phase, consisting of the fat globules, in the cream becomes the continuous phase in the butter; the whey, which is the continuous phase in the cream, also becomes the discontinuous phase in the butter. The churning temperature and time depend on the condition of the cream (Valsechi, 2001).

In general, the time varies from 20 to 40 minutes and the temperature from 10 to 14° C. If the cream is churned at temperatures below 10° C, no butter will form and at temperatures above 15° C, a paste will form from which it is very difficult to separate the buttermilk. The churning time is determined by the size of the agglomerates formed, by noise or by looking through the glass of the churn inspection window (Salinas, 2002).

Excessive churning time will lead to a whey taste in the butter, which will also be short-lived. It is therefore very important to determine the end time of churning. Churns are generally equipped with tubes for washing the fat granules and cleaning the equipment (Clemente and Abreu, 2008).

2.1.4.1.7- Separating the buttermilk

When the fat granules rise to the surface after churning, the buttermilk or butter whey is removed from the bottom of the churn through a screen. Buttermilk should have a maximum fat content of 0.6%. A higher amount of fat in the buttermilk indicates that fermentation has not been carried out properly or the churning has been done incorrectly. Fresh buttermilk has a pleasant taste and is consumed as a refreshing drink. It is also an excellent animal feed (Valsechi, 2001).

2.1.3.1.8- Washing the *butter*

Once the buttermilk has been separated, the butter must be washed to remove the residual whey. This operation is carried out by introducing a volume of water into the churn, approximately equal to the amount of buttermilk removed, and turning the churn a few times at the same speed used for churning. This operation is repeated more than two or three times, or until the wash water comes out clear. The water used for washing must be chemically and bacteriologically pure, and at a lower temperature than that used for churning, to prevent the fat from melting, which would lead to poor washing and a loss of yield (Salinas, 2002).

Normally, washing begins with water at 8° C and at the end, water at 4° C is used, to facilitate draining and prevent the butter from softening during kneading. The butter must be washed thoroughly, eliminating as much as possible of the non-fatty matter in the buttermilk, which is the source of nutrition for the butter's fermentation germs (Berticelli and Motta, 2011).

2.1.3.1.9- Malaxing or kneading *butter*

It is through this operation that the lumps of fat are brought together, giving the butter homogeneity and elasticity, as well as regulating its water content. Incomplete kneading results in excess water in the butter, which will encourage microbial growth, as well as constituting fraud if this content is higher than legally permitted. On the other hand, excessive kneading results in butter with a sebaceous appearance and excessive elimination of water, causing a loss in yield. Malaxing time depends fundamentally on the temperature of the butter. If the temperature is high, the butter will become soft and stick to the walls; if the temperature is too low, it will be difficult for the fat grains to agglomerate. The temperature should be around 12 to 14° C for approximately 10 min (Roman, 2011).

2.1.3.1.10- Salting butter

Normally, butter is salted immediately after churning, when the last of the washing water has been used up. Salt can be added in the form of brine or dry salt. The most efficient process is to use chemically and bacteriologically pure dry salt, in a proportion of 2 to 6%. In this case, approximately 50% of the added salt comes out with the water during malaxing. The salt is added to the butter and you have to wait around 15 minutes before you start kneading, which in this case should take around 20 minutes. It should be noted that in salted butter, the distribution of water is more difficult, resulting in a longer kneading time. As well as giving the butter its flavor, salt also acts as an antiseptic. The amount of salt depends on legal provisions and consumer preference (Valsechi, 2001; Ordóñez *et al.*, 2005).

2.1.3.1.11- Packaging

Butter must be packaged according to its use. Whatever packaging is used, it must meet certain basic requirements:

> - Be impermeable to water vapor and air;
> - Protect the product from sunlight, bacterial contamination and foreign aromas;
> - Facilitate marketing and storage;
> - Encourage consumers to buy them because of their appearance and ease of handling.

The main materials used in butter packaging are aluminum foil and plastic. Paper, usually sulphurized, is permeable to water vapour, air and light. Aluminum is less permeable than paper, and the plastics used are PVC and polyethylene, which have advantages over paper and aluminum (Roman, 2011).

According to Berticelli and Motta (2011), when butter is to be stored or consumed, the larger packages are made of cardboard, lined inside with sulphurized paper or plastic. Cans are also used for packaging butter, in which case they must be perfectly tinned. When filling the packages, air bubbles must be avoided, which would be a point of alteration to the product. The packages are filled by automatic fillers with high precision and capacity.

2.1.3.1.12- Storage

Butter should be stored in cold rooms at temperatures that depend on the length of time it will be stored. If the product is to be consumed quickly, it can be stored between 0 and 7 °C. In the case of prolonged storage, a temperature of between -10 and -15 °C is recommended, which allows for a storage period of many months. The storage chamber must be specifically designed for this purpose and must not contain other products that could easily impart strange aromas to it (ANIL, 2002).

2.1.3.2- Classification

The classification of butter is related to the quality of the raw material, as well as the transformations that take place during its processing, which result in products with different compositions, as shown in Table 2.

Table 2 - Classification of butter according to its characteristics

Features	Extra	1st Quality	Common
Fat	> 83,0	> 80,0%	> 80,0%
Acidity	< 3,0	< 8,0	< 10,0
Salt	< 2,0	< 2,5	< 6,0
Vegetable dye	absent	optional	mandatory

Source: Valsechi (2001).

<u>**III- MATERIAL AND METHODS**</u>

This work was carried out in the provinces of Luanda, Huíla and Cuanza-Sul, specifically in the municipalities of Viana, Cela and Lubango: the characteristics of the localities are detailed below.

3.1- Characterization of the study area

3.1.1- Luanda

Company **A** is located in the municipality of Viana, bordered to the north by Cacuaco, to the east by Ícolo and Bengo, to the south by Kissama and to the west by the Atlantic Ocean and the municipalities of Samba, Kilamba Kiaxi and Rangel (Info-Angola 2015).

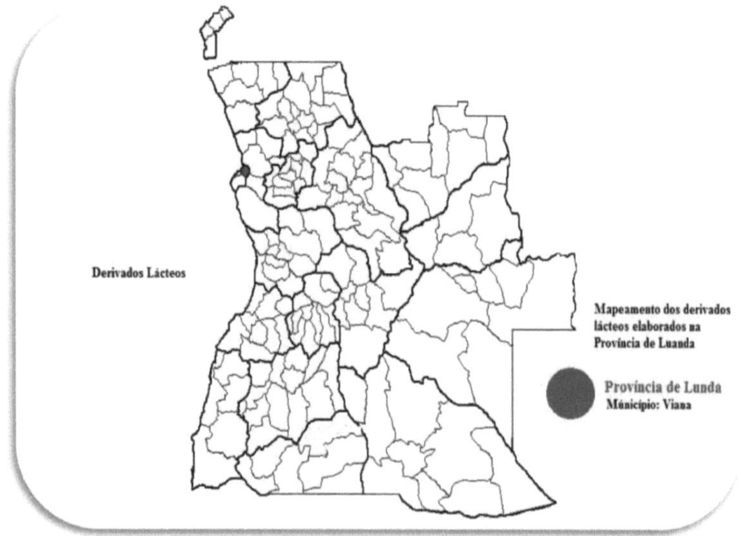

Figure 1: Location map of Company **A,** a producer of dairy products in Luanda province

Source: Info-Angola (2015)

3.1.2- Kwanza-Sul

Empresa de Lacticínios **B** is located in the seat of the municipality of Cela, in the town of Waku Kungo. It is bordered to the north by Quibala, to the east by Andulo, to the south by Bailundo and Cassongue, and to the west by the municipalities of Seles and Ebo (Info-Angola, 2015).

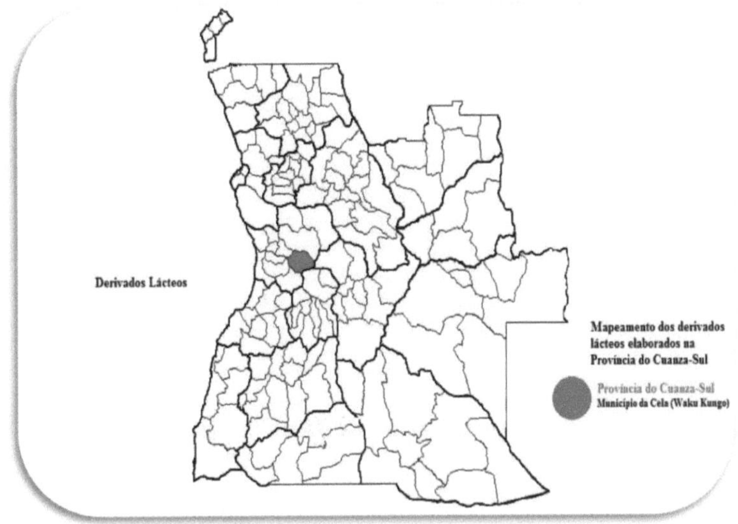

Figure 2: Location map of Company **B,** which produces dairy products in Kwanza-Sul province

Source: Info-Angola (2015)

3.1.3- Huíla

Company **C** is located in the municipality of Lubango in the south of Angola, on the Huíla plateau. It is bordered to the north by the municipality of Quilengues, to the east by the municipality of Cacula, to the south by the municipalities of Chibia and Humpata and to the west by the municipality of Bibala (Info-Angola 2015).

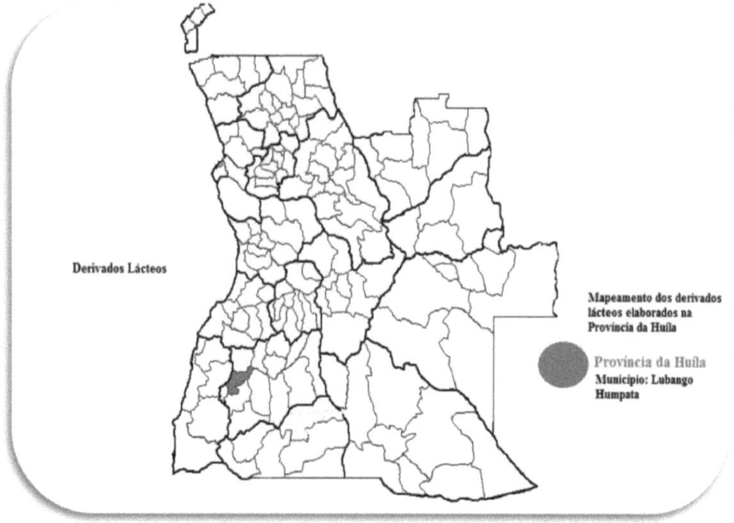

Figure 3: Location map of Company **C,** which produces dairy products in Huila province

Source: Info-Angola (2015)

3.2- Material

- ➢ Yogurt

- ➢ Cheese

- ➢ Butter

- ➢ Thermal box

- ➢ Analytical balance

- ➢ Pipettes

- ➢ Metal tweezers

- ➢ 105 ºC oven

- ➢ Mufla 550 ºC

- ➢ Centrifuge

- ➢ Butyrometer

- ➢ Desiccator

- ➢ Isoamyl alcohol

- ➢ Phenolphthalein

- ➢ Sodium hydroxide

- ➢ Sulphuric acid

- ➢ Petroleum ether

- ➢ Nessler

- ➢ Sodium potassium tartrate

- ➢ Selenium

- ➢ Buffer solutions of pH 4, 7 and 10

- ➢ Computer

3.3- Methodology

The research itself took place between February 2013 and October 2015
and was divided into two stages:

⇨ The first consisted of carrying out a survey with the aim of characterizing Angola's agricultural production chains, listing the plant and animal species used for food production, as well as identifying the agricultural production systems and their productivity, the harvesting and post-harvesting systems used in the communities and the marketing circuits for fresh products and their derivatives. The survey was carried out in the country's 18 provinces by students on the master's degree course in food production and technology, and the location of the dairy products found in the first part of the survey was mapped to facilitate the collection of data and samples to carry out the analyses required for each product, as shown in Annex A.

⇨ The second was based on collecting data and samples to describe the processing of the dairy products produced in the dairy units, determining the physicochemical indicators and statistical processing.

3.3.1-Assessment of the situation of dairy processing plants

In order to achieve one of the recommended goals/purposes, all the steps of the technological manufacturing process were followed:

➡ Monitoring the entire manufacturing process of dairy products, from receiving the raw materials to making the final product. They were processed in different ways, according to the characteristics of each production unit (industrial sector), as can be seen in Annex B.

➡ Physico-chemical analysis of dairy products, and use of information from all the data records provided by the aforementioned companies.

➡ Interviews with company directors, quality control managers and production technicians to obtain information on production, marketing, raw material suppliers, major consumers, difficulties and *marketing*.

The dairies visited carry out control analyses (density, freezing point, boiling point, fat, protein, methylene blue reductase test, alcohol test, antibiotic test) as soon as the raw material arrives, which is one of the main points to be observed and which defines the quality levels of the products to be produced. In addition to verifying the conformity and purity of the product, the scope of the analysis is to know the physicochemical composition.

All the milk was subjected to a pasteurization heat treatment at a temperature of 72.5 °C for 15 seconds according to the methodology described by Ordóñez *et al.* (2005) and had different aspects depending on its purpose, such as yogurt, cheese and butter.

3.3.2- Acquisition of raw materials

With regard to the origin of the raw material, some companies use *in natura* milk from their own units and others, such as company **A**, use powdered milk.

Producers of dairy products are not always producers of fresh milk, but depend directly on the supply of milk from nearby producers. Company **B** not only processes dairy products with its own milk, but also supplies it to other dairy companies. The milk arrives at the production units at a cooling temperature of 3 to 3.5 °C.

3.3.3- *Obtaining the samples*

Different types of yogurt (natural and flavored), butter and cheese were used (3 samples of each product), obtained from the production units, from the companies' marketing sites and from commercial establishments. For each type of product analyzed, samples were acquired randomly and from different batches. Once purchased, the samples were placed in cooler boxes containing ice cubes. The physico-chemical indicators were determined at the Microbiology Laboratories of the Faculty of Agricultural Sciences of the José Eduardo dos Santos University and at the Waku Kungo Veterinary Research Institute in the Food Technology section (Bromatology and Microbiology).

3.3.4- *Determination of physicochemical indicators*

The physicochemical analyses of the samples were carried out in five duly identified replicates and were done in accordance with the analytical techniques described by the Adolfo Lutz Institute (1985), the Adolfo Lutz Institute (2008) and Mexican standards NMF-066-S-1978, NMX-F-083-S-1986. The following parameters were analyzed: pH, total acidity, energy, moisture, fat, protein, ash and carbohydrates.

3.3.4.1- Hydrogen potential (pH)

The pH value was determined by direct reading on a previously calibrated digital potentiometer (PHmeter HANNA HI 221).

3.3.4.2- Total acidity

Acidity was determined by titration with Dornic solution (NaOH) and expressed in °Dornic, where each 0.1 ml of NaOH solution N/9 is equivalent to 1 °D. It was carried out as follows: 1 ml of phenolphthalein was added to 15 ml of the sample diluted with distilled water. This solution was kept stirring as it was titrated with sodium hydroxide (NaOH), until the color changed from white to pink. Finally, the amount of NaOH consumed was recorded.

3.3.4.3- Energy

Energy is expressed in kcal and in kJ/100 g of the edible part. The following indicators were used to calculate it: % carbohydrates $\times$ 4 + % protein $\times$ 4 + % fat $\times$ 9

3.3.4.4- Humidity

4 g of the sample was weighed into a pre-weighed and dried capsule. They were then placed in an oven at 105 °C for 3 hours. They were cooled in a desiccator to room temperature, weighed and the heating and cooling operation repeated until the weight remained constant and the loss of moisture was calculated according to the equation:

$$\% \ \textbf{Humidity} = \frac{(P - P1)}{P2} x100$$

Where:

P= weight of container with wet sample

P1= weight of container with dry sample

P2= weight of wet sample

3.3.4.5- Fat content

To determine the fat content of the yogurt, the Geber method was used: 11 ml of the sample was placed in a butyrometer and then 10 ml of sulphuric acid and 1 ml of isoamyl alcohol were added. The lid of the butyrometer was closed and shaken until the proteins were dissolved, then it was taken to a centrifuge where it remained for 5 minutes and the fat was read directly from the butyrometer.

Two procedures were followed to determine the fat content of the cheese:

⇨ 2 g of the sample was weighed out and a cartridge was made out of filter paper, using the outer part to wrap around the cartridge containing the sample. It was extracted continuously in a Soxhlet apparatus (four to five drops per second), using ether as a solvent; it was kept under heat on an electric plate. The tied filter paper was removed, heated in an oven at 105 °C to remove the remaining solvent, kept for about 1 h; cooled in a desiccator to room temperature; weighed and the procedure repeated until constant weight.

Calculations:

$$\% \ \mathbf{Fat} = \frac{M1 - M2}{M3} x100$$

Where:

M1= mass of the crucible containing the package before extraction.

M2= mass of the crucible containing the package after extraction.

M3= sample mass.

⇨ The amount of fat in the cheese was determined by extraction using the Soxhlet method. The initial weight of the empty extraction cartridge and the cartridge with the sample was recorded, the flask with crystal stones was weighed before extraction, the flask was filled with enough solvent and placed in a water bath at the solvent's boiling temperature for 3 hours. The solvent was extracted using a rotary steamer and the flask was dried in an oven at a temperature of 105 °C; the flask was placed in a desiccator and the flask with the fat and glass stones was weighed.

Calculations:

$$\mathbf{\% \ Fat} = \frac{(m2 - m1)}{M} x100$$

Where:

m1= mass in (g) of the empty round matrix with the glass stones

m2= mass in (g) of the empty round matrix with the glass stones and grease after drying

M = sample weight in g

3.3.4.6- Crude protein

The protein content was determined by weighing 0.1g of the sample, treating it with concentrated sulphuric acid (98%), adding selenium as a catalyst, heating until demineralization, diluting the product in distilled water; the color was developed with 1 ml of Nessler's reagent, adding 2 ml of 50% sodium potassium tartrate. This operation is carried out using a CECIL 1000 SERIES UV: visible spectrophotocolorimeter at a wavelength of 400 nm, using distilled water as a blank test.

The organic matter is decomposed and the existing nitrogen is finally transformed into ammonia. Since the nitrogen content of the different proteins is approximately 16%, the empirical factor 6.38 is introduced to transform the amount of g of nitrogen found into the amount of g of proteins.

Calculations:

$$\textbf{\% Protein} = \frac{a.V.V1.100}{V2.M.1000000} \, x6{,}38$$

Where:

a = amount of N from the calibration curve in µg a g.

M = sample mass.

V = volume of acid consumed

The empirical coefficient in % is considered

3.3.4.7- Ash

To determine the ash content of a foodstuff, the organic matter is calcined after the sample has been carbonized. This was done using the following procedure: 3 g of the sample was placed in previously dried and weighed crucibles. They were then placed on a hotplate to slowly dehydrate them and then placed in a muffle furnace at a temperature of 550 °C for 3 hours for complete incineration until the colored residue turned whitish and ash was obtained. Finally, the same drying, cooling and weighing steps were followed.

Calculations:

$$\% \text{ Ash} = \frac{(P - p)}{M} x100$$

Where:

p - Mass of the empty crucible in grams,

P - Mass of the crucible with ash in grams,

M - Mass of the sample in grams.

3.3.4.8- Carbohydrates

The amount of carbohydrates was determined by subtracting the percentage of moisture, protein, fat and ash from 100, using the following formula: % Carbohydrates = 100 - (% moisture + % protein + % fat + % ash).

3.3.5-Comparison of the physicochemical indicators produced in each industrial sector

The data obtained from the physicochemical determination was compared between the results from the three production units, based on the data established by the international Codex Alimentarius standards.

3.4- Statistical analysis

The data was grouped using the Microsoft Excel spreadsheet version 2013 and submitted to analysis of variance using the Infostat statistical software version 2.00-2013, for the contents of each variable analyzed, where the Tuckey multiple mean comparison test was carried out at a significance level of 5% and the statistical parameters, standard error and coefficient of variation were determined.

IV-RESULTS AND DISCUSSION

4.1- Current situation in the production of dairy products

Figure 4 shows the location map of dairy product production units in Angola, which is the result of surveys carried out in the country's 18 provinces by researchers on the master's course in food production and technology, which took place from January to April 2013, to help identify the regions where dairy products are processed in the country's different production units. They highlighted the provinces of Luanda, Kwanza-Sul and Huila as areas where dairy products are produced.

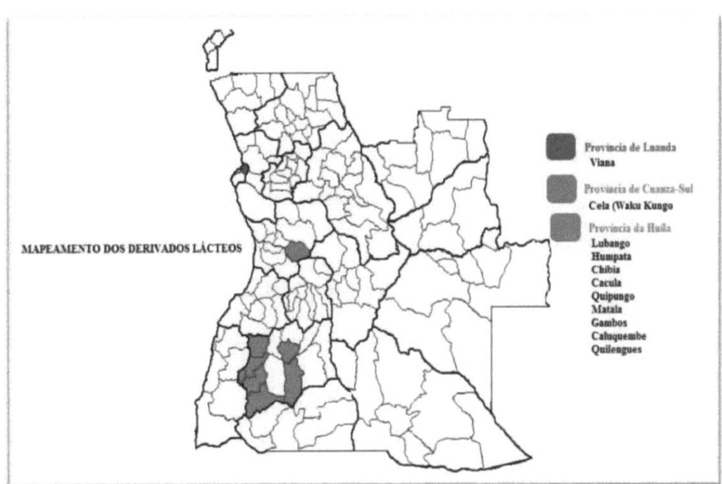

Figure 4: Location map of dairy product production units in Angola

4.1.1 - Production of dairy products at company A

Company **A** is the largest dairy producer in the country. It has a wide production line of UHT milk, kamba milk, condensed milk, chocolate milk, liquid and solid yoghurt in various flavours (banana, pineapple, strawberry, mango, tropical fruit, etc.), dairy desserts and pasteurized butter. They use box packaging made of polyethylene, aluminum for the milks, plastic cups for the yogurts and desserts. Butter is pasteurized and its processing is similar to that of yogurt. It differs only in the addition of agar agar, which makes it thicker.

All the products are sold in the country's various provinces, with the exception of yogurt, which is only sold in the provinces of Benguela, Malange and Cuanza-Norte, due to the fact that the cold chain cannot be broken and due to its very short shelf life.

In order to comply with good production practices, the raw materials are analyzed as soon as they are received, regardless of the fact that some products have a quality certificate containing all the analyses carried out and the processing of the product, as a guarantee of quality production. Therefore, after production, the equipment is mechanically cleaned so that it is ready for the next production.

Much of the raw material they use is bought in Angola and the rest comes from Ireland. They use powdered milk for processing. *In* the past, they used fresh milk (*in natura*) from Company **B,** which didn't meet the company's level of production and they opted to use powdered milk as a raw material. The use of sugar depends on the demands of the market and the product you want to make; in the laboratory, sensitivity and stability tests are carried out on the final product (the production data for dairy products in Luanda province, at company **A**, is shown in Table 3. This is done twice a week for each product). The industry has an annual production of 576,000 L of yogurt and 240,000 kg of butter.

Table 3- Production quantity in 25000 kg of milk

Type of product	Production quantity/liters/kg/day
UHT milk and chocolate	4500
Kamba	6000
Dairy desserts	1000
Liquid yogurt	1000
Solid yogurt	5000
Butter	2500
Cheese	0

This description of Company A's dairy products processing is in line with those described by Silva *et al.* (2010) when they reported that the process of producing dairy products involves a lot of technology, linked not only to the processing units, but also to biotechnology, genetics, biochemistry and chemistry in general. All of these areas are constantly developing due to discoveries derived from research, which is why it is considered an economic milestone.

The same authors state that dairy products are essential in everyone's daily life. This conditions the needs of the market, which leads to the research/creation of new technologies according to the different target audiences. This introduction of new varieties is the main reason for the huge increase in dairy production in recent years. It can therefore be said that this activity has a significant impact on the economic and social spheres.

One of the crucial points regarding social importance relates to the design of new products that take people's needs into account. For example, there are low-fat yogurts, full-fat yogurts, chunky yogurts, flavored yogurts, creamy yogurts and many others, which allows for a greater diversity of choice (Moleta, 2006). This is a reality at Company **A**'s dairy production unit.

*4.1.2 - Production of dairy products in Company **B**'s dairy sector*

In this dairy sector there is a computer room that keeps track of all the animals that go to the milking parlor, as well as the amount of milk the cows provide at each milking.

In the past, milk production was around 26,000 L/month, with an average of 3.2 L cow/day. It is currently producing 60,000 L of milk per month, with an average of 7 to 8 L per cow per day. The company's goal is to achieve an average production of 14 to 16 L per cow/day.

It has a pasteurizer with a capacity of 2,500 L of milk per hour and processes natural and flavored yogurt with different flavors (banana, pineapple, strawberry, mango, passion fruit, among others), ripened and fresh spreadable cheese, butter, desserts and ice cream with different flavors, and recently introduced a production line for sweetened milk and chocolate. The industry's annual production is 720,000 L of fresh milk, 268,800 L of yogurt, 85,728 kg of cheese and 5,760 kg of butter.

The amount of *raw* material used averages around 6,000 L of fresh milk per production and 12,000 L per week. Processing takes place twice a week.

For the production of yogurt, they use 3000 L, of which 1000 L for natural yogurt, 1000 L for strawberry, 500 L for banana and 500 L for passion fruit, which is converted into 2800 L of yogurt, producing 2025 500 ml bottles and 1787.5 L for 200 ml bottles. They use yeast from South Africa to process the yogurt.

For the production of cheese, 3000 to 3500 L *of* fresh milk are used and the yield obtained is 893 kg. For the production of butter, 400 to 600 L of cream are used, with a yield of 36 to 60 kg of the final product. The production of pasteurized fresh milk is 100 L per day, of which 40 L is used for sweetened milk and 60 L for chocolate milk.

The labels are made in Israel and placed on the product inside the facility. The laboratory material comes from South Africa.

According to Siqueira and Almeida (2010), the dairy industry has grown a lot over the last decade, especially the cheese and yogurt processing industries. In the state of Minas Gerais, where the production of milk and dairy products is of great importance to the local economy, there are a large number of companies, many of them small and medium-sized, which compete with large companies from other regions.

The same author states that in order to maintain the competitiveness of the dairy industry, it is not enough to increase the efficiency of individual companies. What is needed is efficient integrated management of the entire production chain, which involves not only these companies, but also milk producers, collectors, transporters, distributors of dairy products, points of sale and others. This is the case with Company **B**'s dairy sector, which has been growing considerably, competing with large companies on the national market.

4.1.3 - *Production of dairy products at Empreza C's dairy sector*

The production unit processes yogurt, cheese, butter and cream. They use *fresh* milk (Figure 5 and 6) produced on their premises and it can be deduced that they have enough raw material to cover their production. The animals are milked twice a day and 216 L of milk are extracted from each milking, with an average of 6.9 L per animal and a total daily milk production of 432 L, totaling 12960 L/month.

Figure 5: Machine milking **Figure 6**: Freshly milked milk

Around 550 to 900 L of milk are used to produce cheese, which is transformed into 67 to 84 kg of cheese as a final product. Matured, semi-matured and curd cheeses are processed under various names depending on how they are used. The whey removed from the cheese was used to feed the pigs.

For the production of yogurt, 300 to 330 L of milk were used and 280 L of yogurt were obtained as a totally natural end product, without the addition of flavorings, essential flavorings or preservatives.

In addition to yogurt, cheese and butter (Figures 7, 8, 9 and 10), the industry also produced curd cheese, cream and fresh milk. The cream produced was of two types: fermented and natural. The milk left over from production was placed in a reserve or storage tank where it remained at a temperature of 3 to 3.5 °C. The sector's annual production is 149,760 L of fresh milk, 26,880 L of yogurt, 8,062 kg of cheese and 1,123.2 kg of butter.

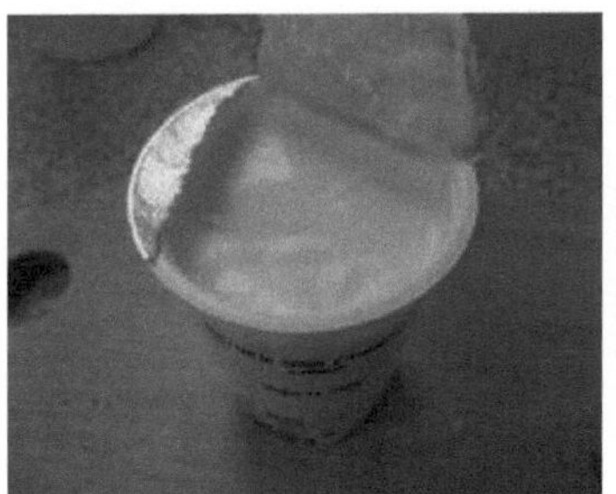

Figure 7: Solid natural yogurt **Figure 8**: Cottage cheese

Figure 9: Cheese   **Figure 10:** Butter

These results are similar to those set out by Nunes (2009) when he reported that the dairy sector is responsible for the production of a wide range of foodstuffs and, in this sense, is commonly accepted as one of the food sectors that involves a broader and more differentiated set of consumers.

According to Siqueira and Almeida (2010) the quality of a dairy product can be observed from the objective perspective; physical, nutritional and hygienic characteristics of the product and from the subjective perspective, i.e. consumer preferences. As such, dairy production is a sector that needs attention at all stages of the production chain, from the raw material to the distribution of the final product. Quality control of a product of this type involves everything from a good laboratory structure,

process equipment and transport vehicles and training for the workforce.

4.2- Description of the stages in the processing of dairy products in the industrial sector

Dairy products are produced in some provinces of Angola, such as Luanda, Huíla, Cuanza-Sul, Benguela, Namibe and Cunene, and are only industrialized in the provinces of Luanda, Huíla and Cuanza-Sul.

4.2.1- Yogurt processing

The freshly milked milk is filtered in order to remove all impurities; it then goes to the pasteurizer at a temperature of 72 ºC for 15 seconds in order to eliminate existing microorganisms; it is then taken to the mixing tank where the culture or yeast is added, remaining for between 5 and 8 hours; then, after fermentation, it is packaged and stored in 200g containers and kept at a temperature of 4 °C.

These yoghurt processing steps are in line with those described by Baptista (2003), Ordóñez *et al.* (2005), Rodrigues, (2007) and Robert (2008) who state that the stages of yoghurt making include: receiving the raw material, mixing and homogenizing, pasteurization, fermentation, cooling, adding the fruit base, packaging and storage.

These results corroborate Giese *et al.* (2010) and Silva *et al.* (2012) when they state that yogurt is the product obtained from bacterial fermentation that leads to increased longevity, bringing numerous nutritional properties.

Yogurt manufacturing flowchart

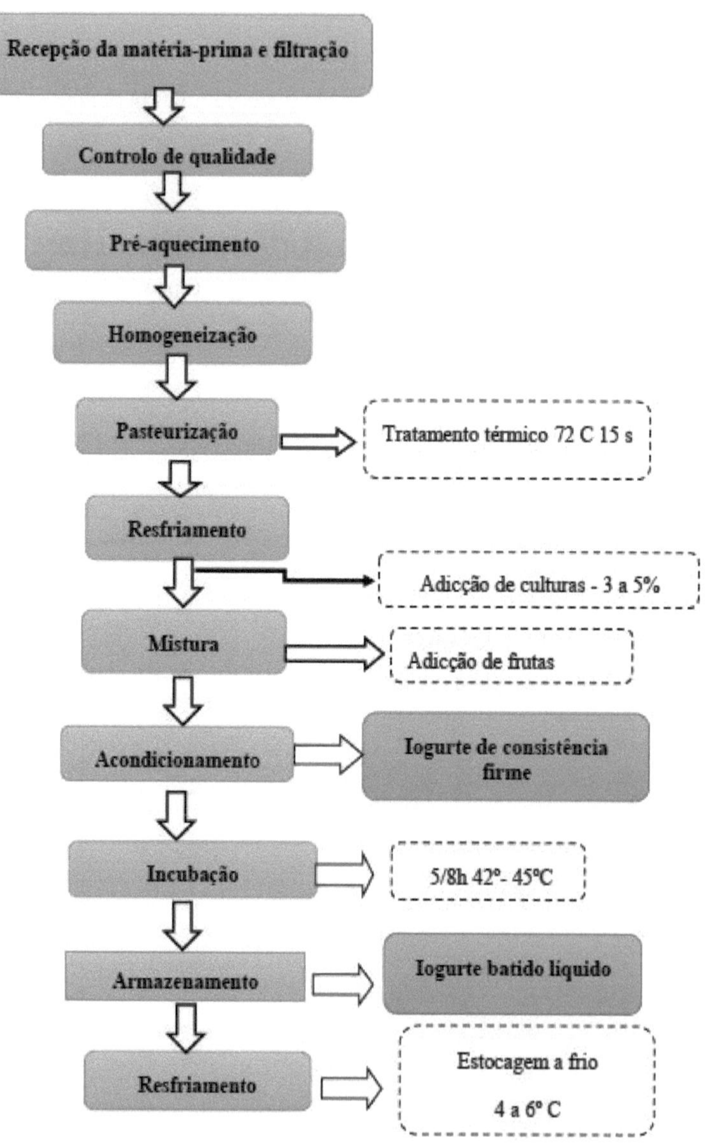

Recepção da matéria-prima e filtração
Controlo de qualidade
Pré-aquecimento
Homogeneização
Pasteurização
Tratamento térmico 72 C 15 s
Resfriamento
Adicção de culturas - 3 a 5%
Mistura
Adicção de frutas
Acondicionamento
Iogurte de consistência firme
Incubação
5/8h 42º- 45ºC
Armazenamento
Iogurte batido líquido
Resfriamento
Estocagem a frio
4 a 6° C

4.2.2 - Cheese processing

After receiving the raw material from milking the animals in the barn, the freshly milked milk is placed in a filter (Figure 11) to remove impurities. A small part of this milk is taken to the centrifuge (Figure 12) where the cream is separated, then pasteurized at a temperature of 72 °C in 15 s and cooled to a temperature of 45 °C.

Figure 11: Filtration of fresh milk   **Figure 12:** Skimming/separating the cream

It goes into the mixing tank where it is heated to a temperature of 32 °C, potassium nitrate, yeast and finally the industrialized rennet powder are added; it is left to mix at the same temperature for 30 min, enough time for it to coagulate (curd formation) (Figure 13). The grains are cut (Figure 14) (0.3 - 0.5 cm^3) and first stirred for 10 to 15 min;

Figure 13: Curd formation.   **Figure 14:** Cutting the curd/1st stirring

Then 17% water is heated to 80 °C and kept at a temperature of 38 °C for 24 min; the second stirring (Figure 15) is done for 20 min; the cheese is separated from the whey (Figure 16) and then placed in the molds for pressing and shaping (Figures 17 and 18). This is done by turning the cheese every 5 min, every 10 min and finally every 20 min; it is left in the molds for 12 h for the molding process (Figure 18); it is dried for a period of 24 h;

Figure 15: 2nd stirring of the curd   **Figure 16:** Desorption

Figure 17: Pressing   **Figure 18:** Molding

It is then placed in the brine, where it remains for a period of 6 to 12 hours depending on the type of cheese you want to produce; it is matured for a minimum of 60 days and then washed with salt until it is completely ripe; finally, it is packaged and stored at a temperature of 4 °C. This process produces half-ripened and ripened cheese.

Serra, Baby Gouda, Gouda, Tisilti, Raccette and Edam are the different types of cheese produced at Company **C**, as can be seen in Table 3: the types of cheese and their respective weights.

The cheese processing stages described above are similar to those described by Valsechi (2001), Perry (2004) Line *et al.* (2005) and Ordóñez (2005), who suggest the following steps: receiving the raw material, pasteurizing, adding the culture, forming the rennet, cutting the rennet, cooking the curd, stirring the curd, desorbing, shaping, pressing, salting and maturing.

The production of cheese is important if we take into account what Farkye (2004) has said: in addition to offering a wide variety of flavors and textures, cheese is a highly nutritious, convenient and versatile food. It is estimated that one third of the world's milk production is used to make cheese.

The European Union (27 Member States) leads the world in cheese production, followed by the United States of America according to Geisler (2011). The main cheese producers in Europe are Germany, France, Italy, the Netherlands and Poland (ANIL, 2011).

Cheese production flowchart

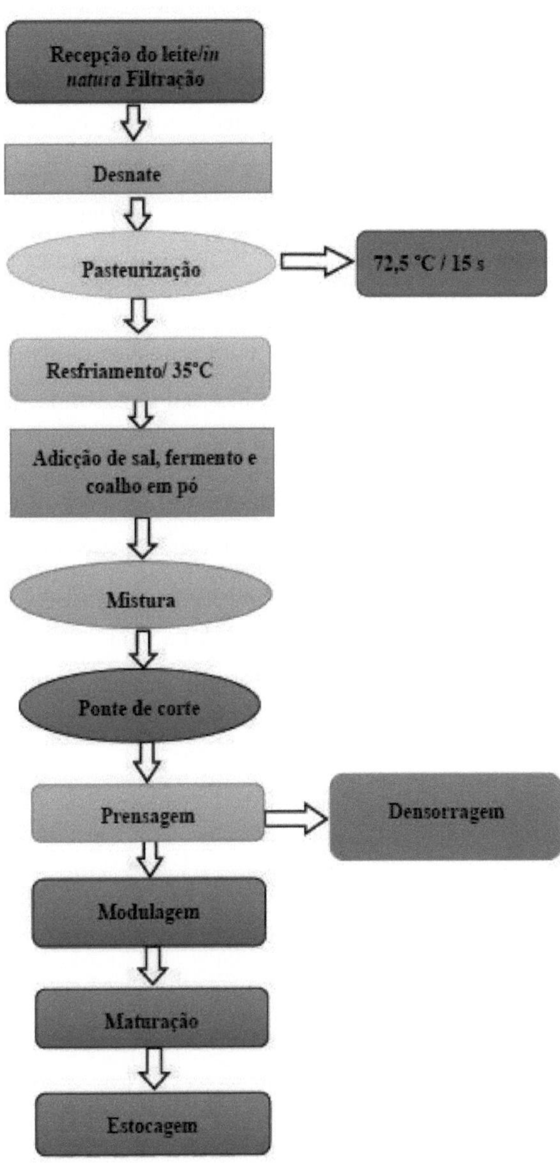

4.2.3 - Butter processing

The production of butter begins with the reception of the raw material (cream) resulting from the centrifugation process at 6000 rpm (separation of cream and skimmed milk); which is then heated in a water bath to a temperature of 82.5 °C and then cooled and taken to the machine where it is beaten until the butter and milk are separated from the butter; the temperature of the water used for washing is 4 °C and it is advisable to repeat the process 5 times; salt is added when the last of the washing water is used up; it is cooled to a temperature of 2 °C; it is packed in plastic containers and aluminum foil. It is vacuum-packed and then sent to the storage room at a temperature of 4°C.

The procedures described here are in line with those of Ordóñez *et al.* (2005), who state that the process of making butter consists of the following phases: receiving the cream, neutralization, pasteurization, deodorization, inoculation with selected cultures, maturation, churning, separation of the whey from the butter or buttermilk, washing, salting, kneading, packaging and storage.

These results are similar to those reported by Carvalho (2008), who states that all these steps result in acidified pasteurized butter. When maturation is not carried out, the final product is pasteurized sweet cream butter.

Flowchart for making butter

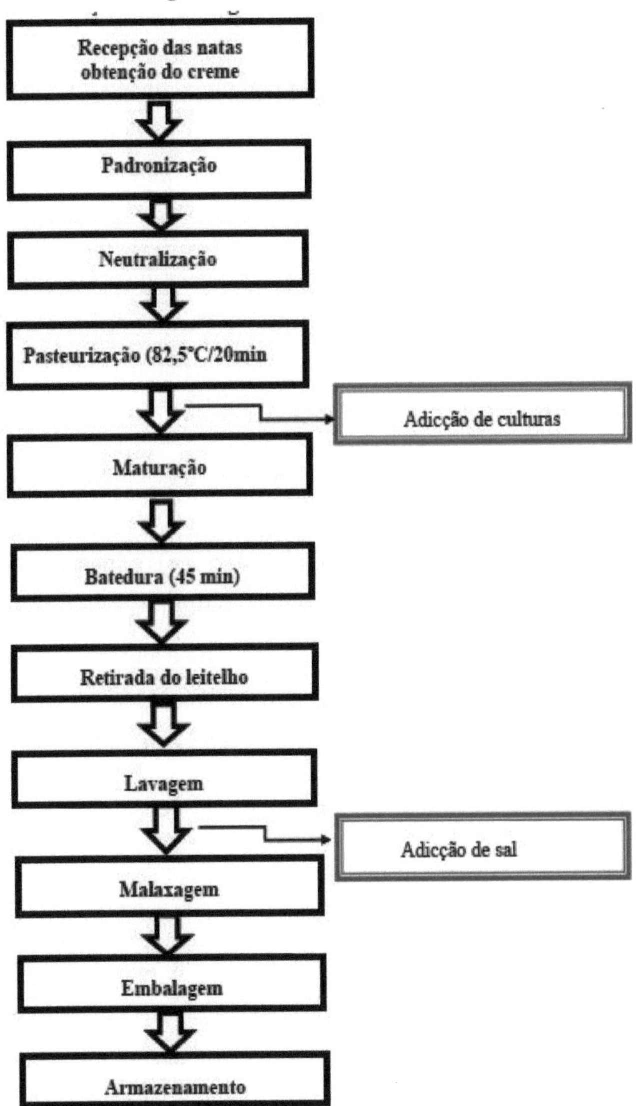

4.3- Cultural aspects of dairy product consumption

In the past, dairy products were simply consumed in the south of the country, where they raised animals that produced milk and processed dairy products at home, as well as being part of the diet of these tribes for cultural reasons. In the north of the country it was not consumed due to the lack of industries to process these products, as well as a suitable form of preservation to reach distant places.

Today, dairy products are consumed in every corner of the country. There is therefore no prejudice against the consumption of yogurt, cheese and butter, as they are all consumed frequently without causing harmful effects due to the quality control of these products.

The people of the South have the custom of consuming milk and its derivatives from cradle to old age. A similar situation was observed by Vidor (2003) when he stated that there are also records of the consumption of fermented milk and curds made from the milk of cows, mares, she-camels and buffaloes by the peoples of North Africa and the Balkans. The fermentation process developed allowed the product to be preserved for a longer period of time, since the processes of preservation by cold or heat were not known.

4.4- Good production practices

The milking parlor at Company **B** (Figures 19 and 20), which supplied milk to the production institutions, has good working conditions and the employees are trained, unlike Company C, where milking takes place outdoors in a containment sleeve.

Figure 19 and 20: Machine milking

Quality milk production is becoming increasingly important. The hygiene of the animal, the milker and the facilities are all necessary to achieve this goal. To ensure proper hygiene, cowherds clean and disinfect the facilities and utensils they use; they disinfect their hands before milking, drying them with paper napkins in some dairies, while in others they use damp towels. The production technicians carry out tests to detect mastitis and the presence of antibiotics to guarantee quality, given that consumers are increasingly demanding a guarantee of the quality of the products they consume.

These procedures are similar to those described by Rodrigues *et al.* (2013) when they state that at the processing or industrial center, the milk is subjected to evaluation tests to verify its quality. Analyses are carried out in accordance with current standards, with the aim of offering the population safe products that do not put consumers' health at risk. It also states that the milker should sanitize their hands and that the cows should be milked with clean, dry teats. To do this, they should wash the teats with chlorinated water, apply an appropriate disinfectant solution to the teats before milking, and dry them with disposable paper towels.

According to the same author, the habit of sanitizing teats after milking is an old one and its dissemination has resulted in a significant reduction in cases of subclinical mastitis in recent years. Dipping the cow's teats in antiseptic is effective after each milking and is the best procedure for reducing the number of bacteria that pass from one cow to another and for reducing the number of infections.

The aforementioned quality control analyses are carried out on the raw material and at every stage of the processing of dairy products, from reception to the final product. For example, on raw milk to check acidity; on pasteurized milk to observe the resistance of microorganisms; on milk with added ingredients to observe the reaction of bacteria; and on cheese whey to check fermentation and acidity.

Similar processes were mentioned by Rodrigues *et al.* (2013) when they stated that laboratory analyses must be carried out in accordance with current legislation, which include: alizarol test, milk acidity, methylene blue reductase test and total bacterial count. The purpose of these analyses is to assess the hygiene and health status of *fresh* milk.

All the production units, in general terms, have an acceptable level of organization for a dairy industry and comply with ISO 9000 and ISO 22000:2005. The facilities for employee use are located within the establishment. Company **C** was found to have animal farms, mainly cattle, close to the processing site.

Wearing uniforms is a habit of the people who process the products, complying with the requirements established by the Codex Alimentarius (2011) and the ISO 22000:2005 standard. As for the processing conditions, the milk is filtered in plastic barrels containing a filter. The milk is mixed and coagulated in a stainless steel tank and the final product is stored at a temperature of 0 to 4 ºC. As for cleaning and sanitizing procedures, in visibly cleaner processing sites, the presence of water, soap and detergent is a constant.

Most of the companies and cheese factories visited have electricity and side roads to access them.

The potential advantages for producers who implement good production practices have been increased competitiveness, the supply of differentiated products and a greater guarantee of permanence in the markets. For consumers, the key is to buy safe, high-quality food.

According to Vargas (2003) and Santos (2007), in recent years several companies and cooperatives in various countries around the world have developed quality programs, which are based on the application of measures on dairy farms to guarantee the safety of dairy products. The implementation of good practice programs in agriculture has been encouraged by various entities and by virtue of legislation in some food industries, a fact which is no exception in Angola.

Thus, good production practices can have a positive impact on obtaining a final product of good health and nutritional quality for the consumer, with a longer shelf life, without losing its physical, chemical and organoleptic characteristics.

4.5- Quantity of fresh milk produced in the production units

Figure 21 shows data on the amount of fresh milk (*in natura*) produced in each dairy production unit each year, where Company **B**'s dairy sector stands out as the largest producer of fresh milk, with a production of 720,000 l/year, followed by Company **C** with 149,760 l/year. These results indicate that Company B's dairy sector is the largest producer of milk because it has the largest number of animals and because of the size and purpose of the sector.

Figure 21: Quantity of fresh milk produced annually in the three production units

According to the FAO (2006), Africa accounts for less than 5% of world milk production, and in most countries production growth is still slow. The main producing countries are Egypt, Kenya, South Africa and Sudan. The same organization states that Egypt is facing production limitations due to restrictions on imports of dairy cattle from countries affected by Enzootic Bovine Leukosis and the drought has affected Kenyan milk production. Africa is one of the continents that has suffered from recurrent drought, and for this reason is expected to increase its total production by only 1%, reaching 36.6 million tons in 2009. A 2% increase is expected in the coming years.

World milk production in recent years is expected to reach 701 million tons, an increase of 1% on previous years. This increase is expected to occur mainly in developing countries, which have been increasing production at a faster rate than developed countries. This difference in production growth will become clear in the coming years, as developing countries are expected to increase production by 4%, while supply in developed countries is expected to remain relatively stable. Overall, world milk production is expected to grow by 2% by 2015, reaching 714 million tons (FAO, 2010).

According to Sequeira *et al.* (2010), world production of cow's milk has gradually increased. It has grown at an average annual rate of 2%, which is higher than the average growth rate of the world population (1.2%) over the last decade. In Asia and Africa, specifically in Angola (in the areas studied with a production of 720,000 l/year in Kwanza-Sul and 149,760 l/year in Huíla), there is a *deficit of* milk and dairy products compared to population demand. The United States has always stood out as the largest producer.

According to Faye and Konuspayeva (2012) *per capita* milk consumption is:

⇨ High (over 150 kg per inhabitant per year) in North America, Argentina, Armenia, Australia, Costa Rica, Europe, Israel, Kyrgyzstan and Pakistan;

⇨ Middle East (30 to 150 kg per inhabitant per year), India, Japan, Kenya, Mexico, Mongolia, New Zealand, the Islamic Republic of Iran, North and South Africa, most of the Middle East and much of Latin America and the Caribbean;

⇨ Low (less than 30 kg *per capita per* year) in Vietnam, Senegal, most of Central Africa and most of East and Southeast Asia.

The same authors state that milk provides only 3% of dietary energy in Asia and Africa, compared to 8 to 9% in Europe and Oceania; 6 to 7% protein in Asia and Africa, compared to 19% in Europe; and 6 to 8% fat in Asia and Africa, compared to Europe, Oceania and the Americas, where it provides between 11 and 14%.

Figure 22 shows the amount of yogurt processed annually by the companies studied (**A**, **B** and **C**). The results show that Company A is the largest yogurt producer with 576,000 l per year. This was followed by **B** with 268,800 l and lastly Queijaria Serra da N'tandavala with 26,880 l. This discrepancy stems from the fact that it has greater production capacity, is located in the country's capital where the population flow is greater and, because it is a renowned company, there is greater demand for its products.

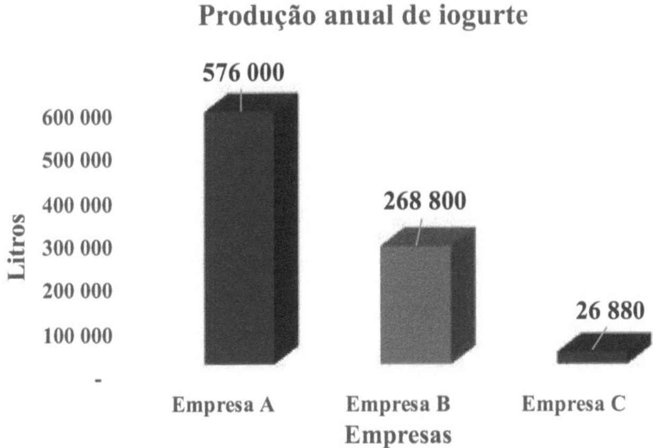

Figure 22: Quantity of yogurt produced annually in the three production units

The province of Luanda, due to its high population density, is a good option for yogurt production, which is in line with Oliveira (1996) when he states that with the growing demand for yogurt by consumers, who have easily incorporated it into their eating habits, the industries have set out in search of new types and flavors, trying to win over the majority of the market. At the same time, they developed processes with lower costs without damaging the quality of the product.

These results differ from those reported by INE (2007), cited by Nunes (2009) with regard to yogurt production, where the author found that it has increased considerably every year in Portugal, resulting in annual growth rates of 16.01%. This increase is due to the change in strategy of multinational companies in the sector, which have relocated production of these products to Spain. This is a strategy that all the companies under study should adopt in order to achieve major goals in yogurt production in Angola.

The consumption of yogurt in Angola has been increasing considerably, due to the fact that it is the most processed and sought-after product among dairy products, which is in line with Forsythe (2002) and Rocha *et al.* (2005) when they state that the demand for and consumption of this product has been intensifying every year and the development of the market is due to the pleasant organoleptic characteristics of the product combined with the nutritional properties resulting from the consumption of yogurt. The taste, texture and aroma of this food is due to microbial metabolism and is considered to be microbiologically safe.

All the production units visited produced yogurts with different flavors, which is in line with Coelho and Rocha (2000) who show that different flavors of yogurt make it possible for people who don't like the taste of milk to consume it. Another factor that contributes to the acceptance of the product, in addition to acidity, is flavoring, which can be done with a wide variety of *fresh* fruits, fruit pulps or juices used in the preparation of yogurt, and this has especially won over consumers who are eager for novelties. Generally, fruits such as strawberries, peaches, plums and coconuts are used, as well as instant coffee and chocolate (Coelho and Rocha 2000; Rocha 2005; Martins *et al.,* 2008).

Figure 23 shows data on the amount of cheese processed annually in the production industries, where Company **B** was found to be the largest cheese producer with 85,728 kg, followed by Company **C** with 8,062 kg per year. Company **B** is the biggest cheese producer because it is medium-sized, has been on the market for a long time, has produced a large quantity of raw materials and its products are known in most of the country's provinces, compared to Company **C,** which is smaller and has been in existence for less time than Company **B,** whose main objective is to produce cheese.

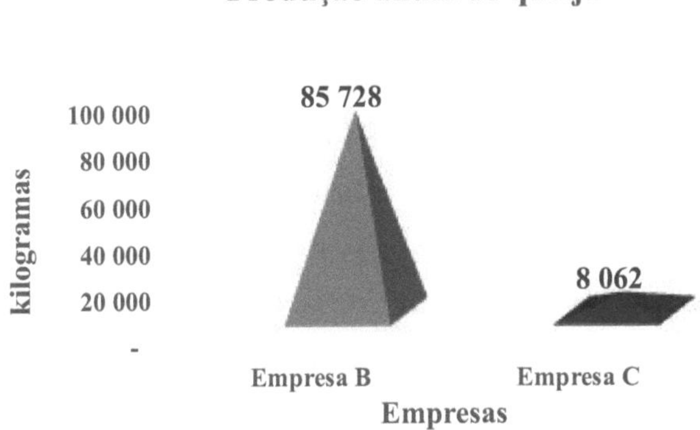

Figure 23: Quantity of cheese produced annually in the production units.

According to figures released by the OECD/FAO, world cheese production in 2010 stood at around 20.6 million tons, representing a 2% increase on 2009. Forecasts for 2011 show an increase in this figure of around 1.8% compared to 2010, and it is currently estimated that production will increase by around 15% in 2018 (TodoAgro, 2011). What's more, many cheeses are produced on a small scale and the great diversity of technological processes used in their manufacture results in different physical, chemical and microbiological characteristics of the processed product.

The results of this study indicate that production in the units studied is very low compared to the FAO (2010) reports, which state that cheese production has increased by 19 million tons in recent years and is expected to continue to increase in the coming years due to its high demand. It also states that among the dairy products sold worldwide, cheese deserves to be highlighted as it accounts for around 40% of the total value of exports, which is not the case in Angola.

The same organization states that the largest producers of cheese in the world per year are: United States with 4461.00 T, Germany 1940.75 T, France 1870.23 T, Italy 1161.49 T, Holland 732.52 T, Brazil 640.00 T, Poland 594.33 T, Argentina 515.00 T, Egypt 462.00 T, Russia 430.00 T, United Kingdom, 343.00 T, Australia, 335.00 T and New Zealand with 329.00 T. As can be seen from the results of this study, it is clear that Angola is still a long way from being on this list, but as long as it works hard, it is possible, in the medium term, to occupy a prominent place by increasing production.

Figure 24 illustrates the amount of butter processed annually in the three production units. Company **A** is the largest butter producer in the country with 97% (240,000 kg), followed by Company **B** with 2% (5,760 kg) and finally Company **C** with 1% (1,123.2 kg) per year. These differences in data are due to the same factors mentioned above. In addition, Company **C** produces minimal quantities because only a small part of its milk production is used to make butter, which is not the target product of the dairy.

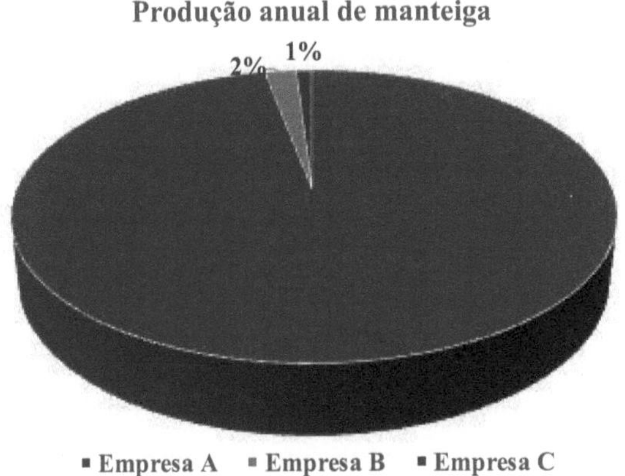

Figure 24: Quantity of butter produced annually at the three production units.

There is a lack of data on butter production in Angola, so our results will be discussed in relation to work carried out in Europe, with emphasis on France, which is considered to be the home of butter, due to its high production and consumption rates.

Angola needs to increase butter production as it is part of people's daily lives. As Valsecchi (2001) reports, butter is one of the most notable ingredients in world cuisine due to its versatility in the kitchen and its presence on people's tables from breakfast onwards. The home of butter is undoubtedly France, where it is consumed at a rate of 17 kg/year/inhabitant. People's passion for butter is justified by the fact that it is not only a very tasty product, but also enhances the taste of any dish. Continuous butter production is all the more economically advantageous the larger it is. The butter yield is very good, as the loss of fat in the buttermilk is between 0.4 and 0.5%.

According to Nunes (2009), Portugal is one of the countries with the lowest production of this dairy product; its production doesn't reach 50,000 tons. At the top of the list of largest producers are France and Germany, which are the only countries with production above 250,000 tons.

Figure 25 shows the most produced dairy products in the country, with yogurt being the most produced dairy product with a total of 72%. This is followed by butter with 20%, and finally cheese with 8%. These results indicate that yogurt is produced the most because it is a product that is highly sought after and sold in Angolan markets and supermarkets, and because it is part of the diet of all age groups, especially schoolchildren.

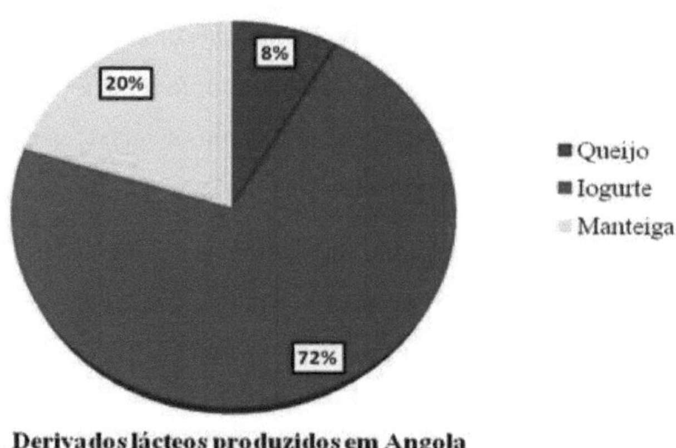

Figure 25: Most produced dairy products in the three production units

In Angola, dairy products play a prominent role in dietary habits, not only because they are considered milk substitutes, but also because they are an important source of calcium and have various nutritional characteristics that are essential for well-being.

The production of yogurt in Angola should increase more and more as it is a food that is on the table of many Angolans, agreeing with Martins (2013) who reports that yogurt is obtained through fermentation, traditionally present on the table of consumers around the world, and is the most produced product in terms of dairy products.

The results obtained in this study are not in line with those reported by Guimarães *et al.* (2008), who state that cheese is the second largest product in the dairy market. Therefore, in our results, cheese is the third most elaborated product. The largest exporter of cheese is Germany, both in terms of product quantity and total economic value.

As Angola is part of the globe, many situations are circumstantial, as Martins (2007) found when he stated that world demand for food has been growing steadily in recent years and in the specific case of dairy products, there has been rapid development. In high-income countries *per capita* consumption is growing at a rate of 2% a year, mainly due to high value-added products. In Eastern Europe, *per capita* consumption of dairy products is 15 kg per inhabitant per year, in Japan 7 kg and in the USA 5 kg, while in Latin America and Africa it is 2 and 3 kg respectively. There is therefore an urgent need to increase the production of milk and dairy products in Angola, as it is a fast-growing market. Following the trend in other countries, consumption of dairy products in Angola has also grown substantially in recent years.

According to Fernandez (2015) the largest producers of milk and dairy products in the world are: European Union with 159 billion liters, United States 93 billion liters, India 58 billion liters, China 42 billion liters, Russia 31.6 billion liters, Brazil 29.7 billion liters, New Zealand 21.5 billion liters, Mexico 11.4 billion liters, Argentina 11.2 billion liters, Ukraine 1.2 billion liters, Australia 9.5 billion liters, Canada 8.7 billion liters and Japan7.3 billion liters of milk per year.

According to the USDA (2010), 165 million tons of milk and dairy products have been consumed worldwide in recent years, which represents an average consumption of 24.5 kg *per capita*. Therefore, when it comes to consumption of milk and dairy products, the figures found in most countries are still below the levels recommended by the FAO, which are 256 l/year for children and 183 l/year for adults.

The outlook for the coming decades is good for all dairy products, as import and export levels are set to increase in practically all countries. Transactions of whole milk and skimmed milk are expected to increase by 8% and 3%, while the forecasts for cheese and butter are 2.4% and 5% respectively.

4.6 - Physico-chemical analysis

Food must comply with the parameters established by the Codex Alimentarius, so that it can be considered fit for human consumption, and so that it is not a means of transmitting agents that could result in damage to consumers' health. It is also imperative to assess their physical and chemical quality, so that consumers are not misled when choosing certain types of food.

Table 5 shows the results for the average pH, acidity, energy, moisture, fat, protein and ash content of the natural yoghurt produced by Company **A**, **B** and **C** in the provinces of Huíla, Luanda and Cuanza-Sul.

Table 5 - Physico-chemical composition of natural yogurt processed at dairy companies **A**, **B** and **C**

Produc tion Unit	n	pH	Ac °D	EN (Kcal)	H0 %	GD %	Ptn %	CO H	Cz %
C	1 5	4,2 5[a]	82,0[b]	40,96, 0[c]	89,4 0[a]	1,0[c]	3,0[b]	4,99 [b]	1,61 [a]
A	1 5	4,4 1[a]	80,0[c]	58,52[a]	86,4 9[b]	1,8[a]	3,83 [a]	6,75 [a]	1,13 [a]
B	1 5	4,3 4[a]	84,0[a]	56,01[b]	86,3 7[b]	1,25[b]	3,0[b]	8,19 [a]	1,19 [a]
EP		0,0 2	1,15	0,44	0,41	0,01	0,03	0,04	0,00 3
CV%		3, 85	2,15	2,64	1,8	0,7	5,08	2,40	7,03

Different letters in the same column indicate significant differences (p≤ 0.05)

EP- Standard error; **CV-** Coefficient of variation; **n-** number of samples and repetitions.
Ac- Acidity; **EN-** Energy; **H0-** Moisture; **GD-** Fat; **Ptn-** Protein; **COH-** Carbohydrates; **Cz-** Ash.

In terms of pH and ash, there was no significant difference between the natural yogurts produced at the three units. However, in terms of acidity, there was a significant difference between the three units, where the natural yogurt processed at company **C** significantly ($p \leq 0.05$) exceeded that of companies **A** and **B**. These results are within the range of parameters established by the Codex Alimentarius (2011), CODEX STAN 243-2003, which establishes pH levels of 4.5 to 4.6 and acidity levels of 60 to 150 °D.

These results are similar to those described by Souza (1991), who states that the ideal acidity range is between 0.7 and 0.9% lactic acid, although the most common values are between 0.7 and 1.25% lactic acid. However, the results of the acidity analyses found in Table 5 were within this range. The same author reports that the determination of acidity can provide valuable data in assessing the state of preservation of a food product.

According to Rodas (2001), it is important to maintain the right balance of bacteria so that the product remains sufficiently acidic and aromatic. Acidity makes yogurt a relatively stable food by inhibiting the growth of Gram-negative bacteria, and the pH can vary from 3.6 to 4.2 and can reach 4.5.

There was a significant difference ($p \leq 0.05$) between the energy value of the different yogurts produced at the three units. The yogurt from Company **A** significantly exceeded the energy value of the yogurt from Company **B** and **C**.

The yogurt made by Company **C** had a significantly higher moisture content ($p \leq 0.05$) than that of Companies **B** and **A**.

As for the fat content, there was a significant difference between the three production units, where the yogurt from sector **A** significantly exceeded (p≤ 0.05) the fat value of the yogurt from sectors **B** and **C**. This result indicates that sector **A** processes yogurt with semi-skimmed milk. Company **C, on** the other hand, produces its yogurt with skimmed milk (low-fat milk) and therefore has less fat. They are within the parameters established by the Codex Alimentarius (2011) for fat content CODEX STAN 253-2006.

The protein content of yogurt from Company **A** was significantly higher (p≤ 0.05) than the protein content of yogurt from Companies **B** and **C**.

These results are similar to those reported by Oliveira *et al.* (2012), in terms of protein, which was 3.72%, and higher in terms of fat, 0.02%, and ash, 0.70%.

The results of the protein and fat content of the yogurts are in line with the legislation in force in Brazil, according to MAPA (2000), which establishes a minimum of 2.9% milk protein. With regard to the fat content, the resolution establishes a range of 3.0 to 5.9% for full-fat yogurts, 0.6 to 2.9% for semi-skimmed yogurts and a maximum of 0.5% for skimmed yogurts.

Our results are close to the values established by Zumbado (2002) in his description of the composition of the nutritional table in g/100g for natural yogurt with the following parameters: energy 53 kcal, moisture 89%, proteins 3%, fat 2.4% carbohydrates, 4.9% and ashes 0.7% and INSA (2013) with the following values: energy 54 kcal, moisture 87.9%, proteins 4.2%, fat 1.8% carbohydrates, 5% and ashes 0.75%.

Table 6 shows the pH, acidity, energy, moisture, fat, protein and ash values of the flavored liquid yogurt made at the Aldeia Nova and Lactiangol dairies. As can be seen, there was no significant difference between the pH, moisture, fat and carbohydrates of the yogurt made at the two production units.

There was a significant difference ($p \leq 0.05$) in the protein, ash and acidity content of Company **B**'s flavored yogurt compared to **B**'s yogurt. The energy value of Company **A**'s flavored yogurt was significantly higher than that of Company **B**. The parameters analyzed are within those established by the General Standard for the Use of Dairy Parameters (CODEX STAN 206-1999).

Table 6- Physico-chemical composition of flavored liquid yogurt processed in the dairy sector (**A** and **B**)

Production Unit	n	pH	Ac ° D	EN (Kcal)	H0 %	GD %	Ptn %	CO H	Cz %
B	15	4,23[a]	75,0[a]	86,85[b]	79,0[a]	1,45[a]	3,0[a]	15,45[a]	1,1[a]
A	15	4,40[a]	56,0[b]	87,12[a]	78,7[a]	1,5[a]	2,8[b]	16,18[a]	0,82[b]
EP		0,03	0,20	0,27	0,48	0,1	0,1	0,37	0,08
CV%		4,14	9,5	4,15	0,94	0,88	4,87	3,29	12,2

Different letters in the same column indicate significant differences ($p \leq 0.05$)

EP- Standard error; **CV-** Coefficient of variation; **n-** number of samples and repetitions.
EP- Standard error; **CV-** Coefficient of variation; **n-** number of samples and repetitions.

Ac- Acidity; **EN-** Energy; **H0-** Moisture; **GD-** Fat; **Ptn-** Protein; **COH-** Carbohydrates; Cz- **Ash**.

Our results were similar to those obtained by Braga *et al.* (2012) in their work carried out in Campina Grande where the physicochemical characterization of yogurts with added fruit pulp and artificial syrup indicated that moisture (76.24 and 78.80%), ash (0.88 and 0.82%) and fat (1.8 and 1.1%) varied from one product to another.

The pH values obtained in this study were higher than those found by Silva (2012) when analyzing the pH of industrialized yogurts, which ranged from 3.72 to 3.86 and were similar to those found by Brandão (1995), which ranged from 4.2 to 4.4.

With regard to the acidity of the products analyzed, the acidity value was 56 °D for Lactiangol yogurt and 75 °D for Company **A** yogurt. According to these results, the acidity values of Company **A**'s yogurt are below the ideal acidity range described by MAPA (2002), which is 0.6 to 1.5% lactic acid. The acidity value of Company **B**'s yogurt is within the parameters established by Codex Alimentarius (2011).

According to Moreira *et al.* (1999), when they analyzed three batches of four yogurt brands, they found that the result found for acidity was not constant across the series of samples, depending on the three manufacturing batches. In general terms, the acidity content was 1.0% and the desirable value would be around 0.70-0.72% lactic acid.

For fat, the values obtained (1.45; 1.5%) were similar for Company **A** and **B**, despite being processed with different raw materials, fresh and powdered milk. This means that the milk used to process them was semi-skimmed milk, which has a maximum fat parameter of 2.9%.

The results presented here are similar to the values established by Zumbado (2002) when describing the composition of the nutritional table in g/100, g for flavored yogurt with the following parameters: energy 86 kcal, moisture 80.4%, protein 1.8%, fat 2.5% carbohydrates 14.1% and ash 1.2% and INSA (2013) with the following parameters: energy 70 kcal, moisture 83.3%, protein 3%, fat 1.3% carbohydrates 11.5% and ash 0.75%.

From a nutritional point of view, yogurt retains all the properties of milk, but there is a slight loss of some water-soluble vitamins due to heating. The amount of fat in milk varies according to the breed of cattle, and the industries are subject to standardization established by legislation, which recommends a minimum of 3% fat according to (MAPA, 2013).

Table 7 shows the pH, energy, moisture, fat, protein and ash values of the ripened cheese from the three production units where the research was carried out. With regard to pH, there was a significant difference between the cheeses produced at all the units in question, with the cheese produced at Company **A** significantly outperforming the values obtained for the cheese from Company **B** and **C**.

In the case of the moisture content, fat, protein and energy value of the cheese in the dairy sector of the industries, it can be seen that there are differences between the results of the cheese from company **A** and **B**, but they are not significant. However, these results significantly exceed the values obtained in the cheese produced at Company **C**, in terms of fat, protein and energy, and the moisture content of the cheese from Company **C** significantly exceeds the cheese produced at **A** and **B**.

With regard to the ash values obtained, it was observed that the ash from the cheese produced in Aldeia Nova significantly exceeded the ash value of the cheese from Company **A** and **C**. There was a difference between the results from the two production units, but it was not significant.

Table 7- Physico-chemical composition of the cheese processed in the dairy sector of **A**, **B** and **C**

Production Unit	N	pH	EN (Kcal)	H0 %	GD %	Ptn %	CO H	Cz %
C	15	4,72 [b]	304,63 [b]	54,22 [a]	27,63 [b]	12,17 [b]	1,82 [a]	4,16 [b]
A	15	6,47 [a]	404,7 [a]	42,57 [b]	38,30 [a]	14,81 [a]	0,19 [c]	4,13 [b]
B	15	5,20 [b]	415,6 [a]	40,80 [b]	39,68 [a]	14,02 [a]	0,60 [b]	4,9 [a]
EP		0,04	0,8	0,54	0,4	0,2	0,49	0,004
CV%		3,7	1,68	4,16	2,01	3,31	9,77	1,52

Different letters in the same column indicate significant differences ($p \leq 0.05$)

EP- Standard error; **CV**- Coefficient of variation; **n**- number of samples and repetitions.
EP- Standard error; **CV**- Coefficient of variation; **n**- number of samples and repetitions.
EN- Energy; **H0**- Moisture; **GD**- Fat; **Ptn**- Protein; **COH**- Carbohydrates;
Cz- Ash. (page 72).

The results of the survey show that the cheese produced at Company **C** is high humidity, as it falls within the classification scale for cheeses with this characteristic (soft cheese), with humidity between 46.0 and 54.9%. When Company **A** and **B**'s cheese is of medium humidity (generally known as semi-hard cheese): humidity between 36.0 and 45.9%. According to Ordóñez *et al.* (2005) and Bezerra (2010), fat and moisture content can vary depending on the type of cheese.

With regard to fat content, the cheese produced in the 3 production units is classified as semi-fat because it is in the 25.0 to 44.9% fat range, according to FIL 4A: 1982, which is close to the values found by Nassu *et al.* (2006), who states that the variations between cheeses are reflected in the raw material and the processing used. For the same author, the physico-chemical characteristics of rennet cheeses showed great variability between samples, which indicates different forms of processing.

The results for the pH (5.18) and protein (14.91%) of the cheese are similar to those observed by Sacunduinga (2011) when he analyzed the physicochemical characterization of the cheeses produced at the Aldeia Nova Project dairy and differ in terms of moisture (72.8%) and fat (6.42%). The variations in the data can be explained by the fact that the author worked with semi-dry and fresh cheese.

The results obtained through the research carried out are close to the values verified by Freitas *et al.* (2013) in studies carried out in the state of Paraíba where they evaluated cheese from three production units, coding them as A, B and C. In the cheeses, differences were found in at least two of the three samples analyzed, in all the physicochemical analyses. The cheese from producer B had high humidity (59.3%), above the 54.9% maximum value recommended by Brazilian legislation for rennet cheese.

According to the same author, in relation to the fat content in total solids, the average values found for producers A, B and C were 50.40, 47.08 and 45.13% respectively, all of which were classified as full-fat cheeses, meeting the established standard (between 35.0 and 60.0%). The average levels of protein and ash in the samples from the three producers ranged from 17.17 to 22.64% and from 2.88 to 3.54%.

It can therefore be seen that the data presented does not differ from the values established by Zumbado (2002), in his description of the composition of the nutritional table in g/100,g for cheese with the following parameters: energy 377 kcal, moisture 38.8%, protein 23.5%, fat 30.1%, carbohydrates 3.0% and ash 4.6% and are similar to those described by INSA (2013) with the following parameters: energy 389 kcal, moisture 36.5%, protein 25.5%, fat 31.5%, carbohydrates 0.2% and ash 5.30%.

The different types of cheese are related to composition and depend on each manufacturing process. Proteolysis is important in the coagulation of hard and semi-hard cheeses and contributes to the development of flavor and texture. These changes in texture during coagulation and fermentation are the main factors contributing to the quality and acceptability of cheeses (Attaie, 2005; Buriti and Rocha, 2005).

Cheese is recognized in Angola as a first-class food, in line with Lopez *et al.* (2012) who state that the most important value of the food is its high quality protein content. Not only does it have a high calorie content (protein - fat), but it also contains large amounts of calcium, phosphate, vitamin A and riboflavin. Such extraordinary properties make cheese an excellent food for children and adolescents. It is undoubtedly considered the king of foods.

Table 8 shows the pH, moisture, fat, protein and ash levels of the butter produced at the three production units. The data shown for humidity is not within the identity and quality parameters established by international legislation (Codex Alimentarius), which recommends a maximum humidity of 16%, and the values found in this study are very high, ranging from 37.90; 46.87 and 47.30%. These results indicate an excess of water in the butter, which constitutes fraud, as the consumer is being harmed by purchasing more water than recommended by the legislation. It is also indicative of a lack of control in the churning and malaxing stages of the cream.

As for the pH and ash values, there were significant differences between the samples analyzed. The pH value of the butter from production unit **B** significantly exceeded that of the other units. The ash content of the butter from N'tandavala unit **C** was significantly higher than that of Lactiangol and Aldeia Nova, while there were no significant differences between the protein values of the three production units.

Table 8- Physico-chemical composition of butter processed in the dairy sector of production units **A**, **B** and **C**

Production Centers	n	pH	Humidity %	Protein %	Ashes
C	15	$5{,}24^b$	$47{,}30^a$	$1{,}00^a$	$0{,}99^a$

A	15	4,49^c	37,90^a	0,93^a	0,64^c
B	15	5,58^a	46,87^a	1,00^a	0,73^b
EP		**0,09**	**0,98**	**0,01**	**0,10**
CV%		**0,59**	**14,57**	**3,41**	**4,37**

Different letters in the same column indicate significant differences (p≤ 0.05)

EP- Standard error; **CV-** Coefficient of variation; **n-** number of samples and repetitions

H0- Moisture; **GD-** Fat; **Ptn-** Protein; **Cz-** Ash.

According to various studies related to the quality of butter samples, it can be seen that this product presents some alterations that could be harmful to the health of its consumers, as it does not meet the criteria established by the international standards according to Codex Alimentarius (2011) CODEX STAN 279-1971, which could be characterized as fraud or non-observance of good production practices. Therefore, there is a need to assess the quality of the physicochemical and microbiological parameters of the butter consumed by the country's population.

According to the RIISPOA and Ordinance No. 146/96 (MAARA), the physical and chemical parameters that butter must meet are: maximum humidity of 16%, which can be as high as 18%; minimum fat of 82% for unsalted butter and no less than 80% for salted butter; maximum defatted dry extract of 2%; less than 0.5% carbohydrates (lactose) and proteins; 0.15% ash and salt content, which can be added or not, with a maximum of 2% allowed.

According to Cecchi (2003); Araújo *et al.* (2007), cited by Berticelli and Motta, (2011), moisture determination is one of the most important measures used in food analysis. The humidity of a food is related to its stability, quality and composition, and can affect its quality during processing, packaging and storage. The humidity values found by Augusta and Santana (1998), in a study involving butters sold in the city of Rio de Janeiro (RJ), ranged from 11.31 to 19.97% humidity, while those found by Silva *et al.* (2009) ranged from 11.51 to 23.97% in different brands, with 50% of the results of the samples analyzed being at odds with the standard, and 50% of their results being similar to those of the present study (46.66%).

Coelhos *et al.* (2009), when evaluating 50 samples of salted butters from different brands, also found values in disagreement with regard to humidity for some of the samples analyzed. The authors state that excess water can trigger decomposition reactions, especially hydrolytic ones, and encourage the development of deteriorating microorganisms that end up depreciating the product. This was the case with the butter evaluated in this study.

With the exception of the humidity data, which was very high, the other results are close to the values established by Zumbado (2002) in his description of the composition of the nutritional table in g/100, g for butter with the following parameters: humidity 15.1%, protein 1%, fat 83%, ash 0.8% and carbohydrates 0.1%. Similarly, INSA (2013) established the following parameters: moisture 16%, protein 0.1%, fat 1.3%, carbohydrates 0.7% and ash 1.9%.

V - CONCLUSIONS AND RECOMMENDATIONS

5.1 - Conclusions

> The processing of dairy products was similar in the companies studied and Company **A**'s production unit was the largest producer of yogurt and butter and Company **B**'s dairy sector was the largest producer of fresh milk and cheese.

> The dairy products industry complies with the standards for making the products, with the exception of butter, which has a very high moisture content, and natural yogurt with a high ash content, which is outside the parameters established by the Codex Alimentarius General Standard for the use of physicochemical parameters, CODEX STAN 279-1971; 206-1999; 243-2003; STAN 253-2006.

> There was greater stability in terms of quantity (production) and quality (physico-chemical indicators) in the production of yogurt and butter compared to cheese, which was highly variable in the three industries studied.

5.2- Recommendations

> That the Angolan institute for standardization and quality (IANORQ), creates norms and regulations for quality control in the production of milk and its derivatives and supervises their development in all production units in Angola.

> That the dairy products processing industries comply with the quality parameters regarding yogurt ash and butter moisture,

established in the international Codex Alimentarius standards CODEX STAN 279-1971; 206-1999; STAN 279-197.

VI - BIBLIOGRAPHICAL REFERENCES

1. Abreu, L. R. 1997. **Technology and use of milk. Lavras**: FAEPE, p.149.

2. Araújo, P. F. Assis, L. M. ; Medina, A. L ; Zambiazi, R. C. 2007. **Physical and chemical quality of homemade butters.** Food Chemistry Course - Department of Food Science - DCA - FCD/UFPel University Campus.

3. ANIL - National Association of Dairy Industries. 2002. **Food Safety: Search for antibiotic residues** in milk. Available at: <http://www.agroportal.pt/a/2002/anil.htm>. Accessed on: 22 09. 2013.

4. ANIL. National Association of Dairy Industries. 2011. **Germany leads European cheese production.** Accessed on April 24, 2012. Available at: www.anilact.pt/informacao-74/4851-alemanha-lidera-producao de queijoeuropeia; Accessed on: 22.10.2015.

5. Attaie, R. 2005. **Effects of aging on rheological and proteolytic properties of goat milk Jack Cheese produced according to cow milk procedures**. Small Ruminant Research, v. 57, p. 19-29.

6. Augusta, I. M.; Santana, D. M. N. 1998. **Evaluation of the quality of extra type butters sold in the state of Rio de Janeiro**. Ciência e Tecnologia. Food. vol.18 n. 4 Campinas.

7. Baptista, P. Venâncio, A. 2003. **Food safety hazards in food processing.** Forvisão: Portugal. ISBN 972-99099-3-8. 109 pp

8. Behmer, M, L. A. 1999. **Milk Technology: cheese, butter, casein, yogurt, ice cream and facilities.** Production, industrialization, analysis. 13.ed. Revised and updated São Paulo: Nobel.

9. Berticelli, D.; Motta, E. 2011. **Physico-chemical and microbiological characterization of butters sold in Francisco Beltrão, Paraná.** Course conclusion paper (undergraduate) Higher Degree in Food Technology, Federal Technological University of Paraná, Francisco Beltrão.

10. Bezerra, J. R. M. V. 2008. **Manufacturing technology for milk derivatives.** Technical bulletin. Department of Food Engineering. Guarapuava: Unicentro. 2008. 56 p. ISBN 978-85-89346-67-2.

11. Bezerra, M. F. 2010. **Physico-chemical, rheological and sensory characterization of yoghurt obtained by mixing buffalo and goat's milk.** Paper presented to the Postgraduate Program in Chemical Engineering - UFRN, as part of the requirements for obtaining a Master's degree in Chemical Engineering. PPGEQ. RN RN/UF/BCZM CDU 637.146.34 (043.2) 116 pp.

12. Braga, A. C. C; Neto, E. F. A. & Vilhena, M. J. V. 2012. **Preparation and characterization of yogurts with mangosteen pulp and syrup (*Garcinia mangostana* L.**) Revista Brasileira de Produtos Agroindustriais, Campina Grande, v.14, n.1, p.77-8477 ISSN 1517-8595.

13. Brandão, S. C. C. 1995. **Industrial yogurt production technology.** Leite e Derivados magazine. v. 5, n. 25, p. 24-38.

14. Brandão, S. C. C. 2002. **New generations of functional dairy products.** Indústria de Laticínios, Jan/Feb, p. 64-66.

15. Buriti, F. C. A.; Rocha, J. S.; Saad, S. M. I.2005. **Incorporation of *Lactobacillus acidophilus* in Minas fresh cheese and its implications for textural and sensorial properties during storage.** International Dairy Journal, v. 15, p. 1279-1288.

16. Carvalho, R. F. 2008. **Butter making.** Bahia: SBRT/RETEC- IEL. Available at: sbrt.ibict.br/dossie-tecnico/downloadsDT/Mjk1. Accessed on: 20.05. 2013.

17. Cecchi, H. M. 2003. **Theoretical and practical foundations in food analysis.** 2nd edition, Campinas - Ed. da UNICAMP.

18. Cidri, A.; Maia, C.; Silva, E.; Serqueira, S. 2005. Preparation of homemade yogurt and evaluation of its acceptability in relation to industrialized yogurts. Revista. In: **Food Hygiene**, vol. 19, n.131, Rio de Janeiro.

19.Clemente, M. G. Abreu, L. R. 2008. Physical and chemical characterization **and oxidative rancidity** of bottled butter fat. Ciênc. agrotec., Lavras, v. 32, n. 2, p. 493-496.

20.CEDGA. European Commission Directorate-General for Agriculture. 2009. **CAP reform: milk and milk products.** 3.ed: Stella ZERVOUDAKI, EC Europe. 8 pp. Available at: *http://europa.eu.int/comm/dg06/index.htm.* Accessed on: 07.08.2013.

21.*Codex alimentarius* .2011. WHO. FAO. World Health Organization. United Nations Food and Agriculture Organization. **Milk and Milk Products** 2.ed. Rome. ISO 9000. Adopted in 1971. Revised 1999. Amendment 2003, 2006, 2008 and 2010. 267pp. ISBN 978-92-5-306786-2. 265 pp

22.Coelho, D. T.; Rocha, J. A. A. 2000. **Práticas de processamento de produtos de origem animal.** 2. ed. Viçosa: Universidade Federal de Viçosa.

23.Coelho, J. A.; Pereira, M. M. G.; Muratori, M. C. S; Klein, J; Manoel H., Lopes, J. B; Lima, D. C. P; Rufino, Y. A. S. 2009. **Physico-chemical quality of butter sold in Teresina,** Pi. Revista Higiene Alimentar. v. 23, n° 170/171.

24.Corazza, S. 2006. **Beauty yogurt**. Available at <http://www.belezainteligente.com.br. Accessed on: 30. 01. 2013.

25.Corte, F. F. D. 2008. **Development of Frozen Yogurt with functional properties.** Dissertation Submitted to the Master's Degree Program in Food Science and Technology at the Federal University of Santa Maria (UFSM, RS), as a partial requirement for obtaining a Master's Degree in Food Science and Technology. 95 pp.

26.Cardoso, R. R. 2006. **Influence of psychrotrophic microbiota on the yield of Minas Frescal cheese made from milk stored under refrigeration.** Viçosa. MG. Master's dissertation, Agricultural Microbiology. Federal University of Viçosa. 43p.

27.Costa, R. G. B.; Lobato, V. Abreu, L. R. Magalhães, F. A. R. 2004. **Salting cheeses in brine: a review**. Rev. Inst. Latic. "Cândido Tostes". No. 336 to 338, vol. 59: p 41-49. Juiz de Fora.

28.Costa, E. N. 2011. **Influence of heat treatment on fatty acids in bovine milk.** State University of Southwest Bahia (Postgraduate Program in Food Engineering - Area of Concentration in Food Process Engineering). Dissertation. Itapetinga. 47pp.

29.Costa A. S. & Lobato, V. 2009. Evaluation of the presence of antimicrobials residues in Milky **Drink and** UHT Milk for Micro**bial** Inhibition Commercial Test. Rev. Inst. Latic. "Cândido Tostes", Mar/Jun, n° 367/368, 64: p 72-76.

30.EMBRAPA. Agricultural Instrumentation Support, Research and Development Unit. 2007. **Making minas frescal cheese.** Available at: www.cpatu.embrapa.br/eu-quero/2007/. Accessed on: 13. 04. 2013.

31.EMBRAPA. Brazilian Agricultural Research Corporation. 2011. **Production, industrialization and commercialization.** (production) Available at: http://www.cnpgl.embrapa.br/nova/informacoes/estatisticas. Accessed on: 24/09/2013.

32.FAO. (Food and Agriculture Organization of the United Nations). 2006. Perspectivas Alimentarias **Análisis del Mercado Mundial**. Available at <http://faostat.fao.org> Accessed on: 20. 10. 2015.

33.FAO. Food and Agriculture. WHO. World Health Organization. 2010. **Word dairy product production: Trade-exports.** contries by commdisty of the United Nation.

34.Farkye, N. Y. 2004. **Cheese technology**. International Journal of Dairy Technology.v.57. n.2-3. p.91-98.

35.Fenemma, O. R. 2000. **Food chemistry.** Zaragoza: Acribia S. A. Delgado-Vargas, F.; Jimenez, A. R. & Paredes-López, O. 2000. Natural pigments: Carotenoids, Anthocyanins, and Betalains -

Characteristics, Biosynthesis, Processing, and Stability. Critical Reviews in Food Science and Nutrition, 2. ed. v. 3, n. 40, p. 173-289.

36. Faye, B and Konuspayeva, G. 2012. **The sustainability challenge to the dairy sector- The growing importance of non-cattle milk production worldwide.** *International Dairy Journal, 24 (2): 50-56.*

37. Fonseca, L. F. L. 2000. **Basic concepts on milk composition and methods used.** *In*: Fonseca L. F. L. da *et al.* Online course on milk quality. São Paulo: Milkpoint.

38. Fernandes, R. V. B.; Botrel, D. A. ; Rocha, V. R. S. V. V.; Ramires, C. S. 2012. Evaluation of the physical-chemical **parameters of common butter**. Revista Académica. Agrarian and Environmental Sciences. Curitiba, v. 10 n. 2 Apr./Jun. p. 171-176. ISSN: 0103-980X and ISSN 1981-4178.

39. Fernandez, V. L. P. 2015. **Situacion de la leche a nível Mundial.** Comité Nacional Sistema Produto Bovino Leche. U. S. Dairy Export Council- Mexico. 25 pp.

40. Ferreira, V. L. P. 2000. **Sensory analysis. Discriminative and affective tests.** Campinas: Brazilian Society of Food Science and Technology. Quality Series Manual. p. 73-77.

41. Ferreira, C. L. L. F. 2001. **Dairy and fermented products**: Biochemical and technological aspects. Cadernos didáticos. 3rd ed. Viçosa: UFV. n. 43. 112 p.

42. Freitas, W. C; Travassos, A. E. R. ; Maciel, J. F. 2013. **Microbiological and physicochemical evaluation of raw milk and rennet cheese produced in the state of Paraíba.** Revista Brasileira de Produtos Agroindustriais, Campina Grande, v.15, n.1, p.35-42. 35 ISSN 1517-8595.

43. Forsythe, S. J. 2002. **Food safety microbiology.** São Paulo: Artmed.

44.Fox, P. F.; Mcsweeney, P. L. H. 1998. **Dairy Chemistry and Biochemistry.** Published by Blackie Academic & Professional, an imprint of Thomson Science, 2-6 Boundary Row, London SE1 8 UK. First. 478 pp.

45.Fox, P. F.; Guinee, T. P.; Cogan, T.M. & Mcsweeney, P. L. H. 2000. **Fundamentals of cheese science.** New York: Aspen. 587 pp.

46.Furtado, M. M. 2005. **Main cheese problems: causes and prevention.** Fonte Comunicações e Editora. São Paulo, SP, Brazil. 200 p.

47.Garcia, G. A. C. 2007. **Effect of the use of proteolytic enzymes on the ripening of reduced-fat Prato cheese.** Dissertation (master's degree) - Universidade Estadual Paulista, Instituto de Biociências, Letras e Ciências Exatas. São José do Rio Preto 154 pp.

48.Gava, A. J. ; Silva, C. A. B. & Frias, J. R. G. 2008. **Food technology**: Principles and applications. São Paulo: Nobel. P. 214-221. ISBN 9788-85-213-1382-3.

49.Giese, S.; Coelho, S. R. M. ; Téo, C. R. Paz A.;Nóbrega, L.H. P.; Christ, D. Revista Varia Scientia Agrárias. 2010. **Physico-chemical and sensory characterization of yogurts sold in the western region of Paraná.** ARTIGOS & ENSAIOS. v. 01, n. 01, p. 121-129.

50.Geisler, M.. 2011. **Cheese industry profile.** Available at: http://www.agmrc.org/commodities_products/livestock/dairy/cheese_industry_profile cfm; Accessed on: 20. 10 2015.

51.Goetze, M. 2010. **Evaluation of the quality of sheep's milk for the production of pecorino toscano and feta cheeses.** Federal Institute of Education, Science and Technology Rio Grande do Sul. Bento Gonçalves Campus. Course conclusion paper presented to the Higher Degree in Food Technology at the Bento Gonçalves

Campus of the Federal Institute of Education, Science and Technology of Rio Grande do Sul, partial requirement for course conclusion. 54 pp.

52.Gori, A. A. 2012. **Rapid method to discriminate season of production and feeding regimen of butters based on infrared spectroscopy and artificial neural networks.** Journal of Food Engineering, v. 109, p. 525-530.

53.Goulet, J. 1991. **Milk and fermented milk products.** In: Amiot, J. Ciencia y tecnologia de la leche. Zaragosa, Spain: Editorial Acribia, ch.12, p. 359-371.

54.Guimarães, T. F.; Carvalho, G. R; Carneiro, A.V & Duarte, M.M. 2008. **World cheese exports**: 2003 to 2007. X Minas leite.

55.Hartmann, W. 2009. **Physical-chemical, microbiological, management and hygiene characteristics of bovine milk production in the western region of Paraná: occurrence of *Listeria monocytogenes*.** PhD thesis presented to the Postgraduate Program in Food Technology at the Federal University of Paraná. Curitiba. p. 88

56.Hohendorff, C.G. 2006. **Cheese production** p. 4. Available at: http: www.enq.ufsc.br/labs/probução.queijos Accessed on: 25. 03. 2013.

57.IAL. Adolfo Lutz Institute. 1985. **Analytical Standards of the Adolfo Lutz Institute: Chemical and physical methods for food analysis.** 3. ed. São Paulo: IMESP, v. 1. p. 43-44

58.IAL-Instituto Adolfo Lutz -.2008. **Physico-chemical methods for food analysis.** 1. ed. digital. São Paulo.1020.p. Available at: <http://www. ial.sp.gov.br/index.php?optio. Accessed on: 17 09. 2013.

59.Info-Angola. 2015. **Provinces of Angola.** Available at; http//www.infoangola/índex.php. Accessed on: 22.10.2015.

60.INE, I.P. 2007. **Agricultural Portugal 1980-2006,** Lisbon - Portugal. 121 pp. Available at:www.ine.pt/xportal/xmain?xpid=INE&xpgid=ine_publicacoes &PU. Accessed on: 22.10. 2015.

61. ILCT. Cândido Tostes Dairy Institute. 1977. **Handout on cheese-making technology.** Juiz de Fora.

62. INMETRO. National Institute of Metrology, Standardization and Industrial Quality. 2006. **Fresh and standard minas cheese.** Brazil. Available at: http://www.inmetro.gov.br/consumidor/. Accessed on: 24. 04. 2013.

63. INSA. Instituto Nacional de Saúde Doutor, Ricardo Jorge. 2013. ISO54. ISO76. ISO79. IS385. Online. **Food and Nutrition: Food** Composition Table. Portugal. Available at: http://www.insa.pt/sites/INSA/Portugueses/AreasCientificas/Alim entNutricional/Aplicacao/Online/TabelaAlimento. Accessed on: 25.11.2013.

64. IPOQ Portuguese Standard 1711. Alfa-Laval. Dairy Handbook. Available at: http://www.foodsci,uoguedlph.ca/dairyedu/butter.html. Accessed on: 07.04,2014

65. Katiki, L. M. 2004. **Elaboration of mold-ripened cheese obtained from mixed coagulation, with frozen goat's milk and frozen curds.** Paulista State University "Julio de Mesquita Filho" Faculty of Veterinary Medicine and Zootechny Botucatu Campus. Dissertation presented to the Faculty of Veterinary Medicine and Zootechny, UNESP - Botucatu Campus, to obtain a Master's Degree in Veterinary Medicine - Area of Concentration: Health Surveillance.

66. Lidon, F. J. C. & Silvestre, M. M. A S. F. 2007. **Food industries additives and technologies.** Lisbon: Escolar editora. p. 197-211. ISBN. 978-972-592-203-3.

67. Line, V. L. S.; Remondetto, G. E. & Subirade, M. 2005. **Cold gelation of blactoglobulin** oil-in-water emulsions. Food Hydrocolloids, v. 19, p. 269-278.

68.López, M. A. Torres; Lamazares, J. L. Á. & Cañizares, A. V. 2012. **Food hygiene and control. Cuba:** Felix Varela. Habana. p. 125-252. ISBN 978-959-07-1642-3.

69.MAARA- Ministry of Agriculture, Supply and Agrarian Reform. Ordinance No. 146 of March 7, 1996. **Approves microbiological, physical and chemical standards for milk and dairy products.** FIL Standard 50B: 1985. Milk and Milk Products Sampling Methods.

70.MADRP. Ministry of Agriculture, Rural Development and Fisheries. 2007. **Milk and Dairy Products.** Sector diagnosis. document prepared by the planning and policy office. 58 pp.

71.Madeiro, A. P,; Casagrande, F, & Bittarelo, K. P. 2006. **Yogurt**. Federal University of Santa Catarina Eqa - Department of Chemical and Food Engineering, Florianópolis. Available at: http://www.revistalaticinios.com.br/. Accessed on: 28.04.2013.

72.Malta, H. L.; 2001. **Verification of the physical-chemical and microbiological quality of some yogurts sold in the Viçosa region - MG.** Revista do Instituto de Laticínios "Candido Tostes": Anais do XIII Congresso Nacional de Laticínios, Rio de Janeiro, n° 321, p. 152-153.

73.Martins, P. C. A. 2007. **Brazil's competitiveness in the world dairy market.** In: 80 INTERLEITE. Uberlandia. Proceedings. Agripoint. p.13-32.

74.Martins, O. A.; Rudge, A. C.; Meira, D. R. 2008. **Changes in pH, lactic acid and microbiological indicators in different brands of yogurt sold in the city of Botucatu.** São Paulo, Brazil. PUBVET, v. 2, n.19, Art. 224.

75.Martins, G. H.; Kwiatkowski, A; Bracht, L. Srutkoske, C: L. Q.; Haminiuk, Ch. W. I. 2013. **Physicochemical, sensory and rheological profile of yogurt made with water-soluble soy extract and supplemented with inulin.** Revista Brasileira de

Produtos Agroindustriais, Campina Grande, v.15, n.1, p.93-102. 93 ISSN 1517-8595.

76.MAPA- Ministry of Agriculture, Livestock and Supply and Agrarian Reform. 2000. BRAZIL. Ordinance No. 146, of March 7, 1996. **Approves technical regulations on the identity and quality of dairy products.** Diário Oficial da União, Brasília, March 11. 96. Section I. 50 pp.

77.MAPA. Ministry of Agriculture, Livestock and Supply and the Health Surveillance Secretariat. Ministry of Health Brazil. 2002. **New commented legislation on dairy products.** Brasília, DF. 327p.

78.MAPA- Ministry of Agriculture, Livestock and Supply. Technical Regulation on the Identity of Cheese. 2009. FIL Standard 5B: 1986. **Cheese and Processed Cheese Products.** Available at: <http://www.agricultura.gov.br/>. Accessed on: 11. 12. 2013.

79.MAPA. Ministério da Agricultura, Pecuária e Abastecimento. 2013 Resolução n. 4 de 28 de Junho de 2000 do Ministério da Agricultura Pecuária e Abastecimento. **Federal Official Gazette, Executive Branch**. Brasília, DF, July 5, 2000. Section 1, p. 5. Resolution for the denomination of common butter marketed in national territory. BRAZIL.

80.Moleta, C. B. 2006. **Preparation of homemade yogurt and physical-chemical evaluation in relation to industrialized yogurt.** Work presented as a requirement for completion of the degree in Nutrition at the Assis Gurgacz College. Cascavel. 11 pp.

81.Moreira, S. R.; Schwan, R. F. Carvalho, E. P.; Ferreira, C. 1999. **Microbiological and chemical analysis of yogurts sold in Lavras. Minas Gerais** Food Science and Technology (Campinas) Ciências Tecnologia de Alimentos. vol.19 n.1. p. 147-152. Campinas. *On-line version* ISSN 1678-457X Available at: http://www./s_arttextS0101-20611999000100027. Accessed on: 26.12.2013.

82.Muhlbauer, F. B. ; Cesar, G. M.; Junqueira, P. C. L. G.; Gomes, P. C. L.; Souza, A. D. & Furlan, M. R. 2012 **Evaluation of the physical and chemical characteristics of Uvaia pulp and yogurt.** University of Taubaté. Faculdade Integral Cantareira, São Paulo /SP. THESIS, São Paulo, ano IV, n.17. p. 60-77.

83.Mundim, S. A. P. 2008. Preparation of functional yogurt with goat's milk, flavored with fruits from the cerrado and supplemented with inulin. Rio de Janeiro. Dissertation (Master of Science) - Federal University of Rio de Janeiro - UFRJ, School of Chemistry, Postgraduate Program in Chemical and Biochemical Process Technology. 133pp.

84.Nunes, A. S. A. R. 2009. **The milk and milk products sector from a food safety perspective.** New University of Lisbon. Faculty of Science and Technology. Dissertation submitted to the Faculty of Sciences and Technology of the New University of Lisbon to obtain a Master's degree in Food Technology and Safety. 259 pp.

85.Nassu, R.T.; Andrade, A.S.A. De; Silva, A.C. 2006. **Physico-chemical characterization of regional cheeses produced in the state of Rio Grande do Norte**. Revista do Instituto de Laticínios Cândido Tostes, v. 61, p. 303-305.

86.Neves-Souza, R. D.; Silva, R. S. S. F. 2005. **Cost-performance study of Minas frescal cheese processing with soy derivatives and different coagulating agents**. Food Science and Technology. Campinas, v. 25, n. 1, p. 170-174.

87.Mexican standard. NMX-F-066-S-1978. NMX-F-083-S-1986. General Directorate of Standards.

88.NP EN ISO 22000. 2005. Food safety management systems. **Requirement for any organization operating in the food chain;** November 2005. IPQ - Portuguese Quality Institute, Caparica.

89.Oliveira, J. A.; Caruso, B. G. J. ; 1996. **Leite:** Obtenção e qualidade do produto fluido e derivados v.2. Piracicaba - SP: FEALQ.

90.Oliveira. C. 2007. **Milking parlors; Associação dos Jovens Agricultores Micaelenses.** Available at:www. delaval.com.br/Daily-knowledge/Efficient cooling. Accessed on 30. 03. 2013.

91.Oliveira, G. E.; Silva, A. E.; Oliveira, T. E. Froelich, A. 2012. **Physicochemical characterization of homemade butters sold in Alagoas.** 5th National Meeting on Chemical Technology.

92.Ordóñez, J. A. Rodrígues, M. I. C. Álvarez, L. F. Sanz, M. L. G. Minguillón, G. D. G. F; Parales, L. H. & Cortecero, M. D. S 2005. **Food Technology: foods of animal origin.** Porto alegre: Armated. São Paulo. v.2. ISBN 85-363-0431-6. pp. 48-191.

93.Park, Y. K.; Koo, M. H. & Carvalho, P. O. 1997. **Recent advances in functional foods.** Bulletin of the Brazilian Society of Food Science and Technology, Campinas, v. 5. n. 31. p. 200-206.

94.Park, Y.W.; Juárez, M.; Ramos, M.; Haenlein, G.F.W. 2007. **Physico-chemical characteristics of goat and sheep milk.** *Small Ruminant Research*, v.68, n.1-2, p.88-113.

95.Paula, J. C. J.; Carvalho, A. F. C. ; Furtado, M. M. 2009. Basic principles of cheese production: **from historical to salting.** Rev. Inst. Latic. "Cândido Tostes", Mar/Jun, nº 367/368, 64: 19-25.

96.Pernodet, G. 1990. **Technologie comparée des différents types de caillé.** In ECK, A. Le fromage 2° ed. Paris, Lavoisier, part 2, chap.5, p.219-248.

97.Perry, K. S. P. 2004. **Cheese:** chemical, biochemical and microbiological aspects. Química Nova, São Paulo, v.27, n. 2, 293-300.

98.Penna, C. F. A. M. 2011. **Production and quality parameters of milk and cheese from Lacaune, Santa Inês and mixed-breed ewes fed diets containing soy or linseed.** Federal University of Minas Gerais Veterinary School Department of Animal Science. Belo Horizonte. Thesis presented to the Federal University of

Minas Gerais as a partial requirement for the degree of Doctor of Animal Science. Area of concentration: Animal Production.p.154-412.

99.Prudencio, I. D. 2006. **Physical properties of *petit suisse* cheese made with cheese whey retentate and stability of added anthocyanins and betalains.** Dissertation presented to the Postgraduate Program in Food Science at the Federal University of Santa Catarina, as a final requirement for obtaining a Master's Degree in Food Science. 87 p. Florianópolis - SC.

100.Pupin, A. M. 2002. **Probiotics, prebiotics and symbiotics: applications in functional foods.** In: SEMINAR NEW MARKET ALTERNATIVES, Campinas. [Papers presented]. Campinas: ITAL. p. 133-145.

101.Ralph, E. 1998. **Tecnología de los productos lácteos.** 2. ed: Editora Acribia. S. A. Zaragoza-Spain.

102.Reis, B R & Combs, D R. 2003. **Dairying in the United States of America.** Trends in the Consumption of Animal Products in the United States. 10pp.

103.RIISPOA. 1980. **Regulations for the Industrial and Sanitary Inspection of Products of Animal Origin.** Brasília, Ministry of Agriculture.

104.Robinson, R. K. 2002. **Dairy Microbiology Handbook. Third Edition.** ISBN 0- 471-38596- 4. Wiley-Interscience, Inc. New York - USA. Microbiology of soft cheese, 479p 513p. 765 p.

105.Robert, N. F. 2008. **Yogurt making.** D O S S I Ê T E C N I C O. Technology Network of Rio de Janeiro. REDETEC. Available at: Accessed on: 02.10.2013.

106.Rocha, A. M. P. 2004. **Control of fungi during the ripening of standard Minas cheeses,** 96f. Dissertation (Master's Degree in Food Science and Technology) Federal University of Santa Maria - UFSM, Santa Maria-RS.

107.Rocha, E. M. 2005. **Sensory analysis and shelf-life study of tropical fruit-based dairy desserts.** Food Hygiene. São Paulo, v. 19, n. 135, p. 28-33.

108.Rodas, M. A. de B. 2001. **Physicochemical and histological characterization and viability of lactic acid bacteria in fruit yogurts:** Ciência e Tecnologia de Alimentos, Campinas, v. 21, n. 3, p.304-309, 2001.

109.Rodrigues, F. 2007. **Comentariso** IN 46. MAPA (Ministerio da Agricultura Pecuaria e Abastecimento).

110.Rodrigues, E. Castagna, A A. Dias, M T and Aronovich, Marcos. 2013. **Quality of milk and dairy products:** Processes, technological processing and indices. Niterói-RJ: Rio Rural Program. 53 pp. ISSN 1983-5671.

111.Roman, J. A. 2011. **Butter-making technology.** Available at: <http://www.td.utfpr.edu.br/janesca/Manteiga.pdf>. Accessed on: 14. 04 2013.

112. Sacunduinga G. C. CH. 2011. **Characterization of the Technological Process and Physical-Chemical Analysis of the Cheeses** Made at the Dairy Factory of the Aldeia Nova Waku-kungo Project. End of Course Work for the Degree in Veterinary Medicine. P.78.

113.Salinas, R. D. 2002. **Food and nutrition.** 3.ed. Porto Alegre: Artmed. Sawyer, W. H. 1969. Complex between β-lactoglobulin and κ-casein: a review. J. Dairy Sci., v. 52, n. 9, p. 1347-1363.

114.Santos M.V. 2007. **Milk cooling and its impact on quality.** In: Online training: milk quality and milking guidance. Agripoint, Piracicaba- SP. Module 3.

115.Sequeira, K. B; Carneiro, A. V; Almeida, M. F; Nalon, R. C. S. 2010. **The Brazilian dairy market in the global context.** Technical circular. 104. General Mine. Embrapa. *ISSN 1678-07X.* 12 pp.

116. SENAR- National Rural Apprenticeship Service. 2010. **Yogurt, dairy drinks and dulce de leche:** production of dairy products / National Rural Apprenticeship Service. -- 2. ed. Brasília. 68 p. il. I SBN 978-85-7664-047-9. Series CDU 637.1

117. Silva, A. I. D.; Pereira, F. J. C.; Beirão, M. C. R. V. ; Gomes, M. R. F. S. ; Moura, P. C. ; Porfírio, P. A. & Fernandes P. D. L. 2010. **Yogurt Production.** FEUP Project. University of Porto. Faculty of Engineering. 29. pp Available at: http://www.bcsdportugal. org/files/834.pdf. Accessed on: 17. 02. 2013.

118. Silva, L. C.; Machado, T. B. ; Silveira, M. L. R.; Rosa, C. S. & Bertagnolli, S. M. M. 2012. **Microbiological Aspects, pH and Acidity of Homemade Yogurts Compared to Industrialized Yogurts from the Santa Maria Region** Final Undergraduate Paper. Series: Health Sciences, Santa Maria, v. 13, n. 1, p. 111-120. ISSN 21773335.

119. Soroa, J. M. 2001. **Dairy industries.** In: Umbelino, D. C. Cardello, B. A. M. H.; Rosst, A. Aspectos tecnologicos e sensorial do yogurte de soja enriquecido com cálcio, Ciência e Tecnologia de Alimentos, Lisbon: Editora Litexa, 5th ed. v.21, n.3, p. 276-280.

120. Souza, G. 1991. **Yogurt quality factors.** Coletânea do ITAL. v. 21, n.1, p. 20-27.

121. Spreer, E. 1991. **Lactologia Industrial.** 2.ed: Spain: Zaragoza: Acribia. S. A.

122. St-Gelais, D.; Haché, S. 2005. **Effect of b-casein concentration** in cheese milk onrennet coagulation properties, cheese composition and cheese ripening. Food Research International, v. 38, p. 523-531.

123. Tamime, A. Y. 2006. **Brined Cheese.** Lowa: Blackwell Publishing. Available at:http://books.google.com/books. Accessed: 26. 11. 2013.

124. Teixeira, L.V.; Fonseca, L.M. 2008. **Physico-chemical profile of whey from mozzarella and minas-padrão cheeses produced in various regions of the state of Minas Gerais.** Arq. Bras. Med. Vet. Zootec., v. 60, p. 243-250.

125. TodoAgro (2011). **World queso production will increase**. Available   at: http://www.todoagro.com.ar/noticias/nota.asp?nid=14649; Accessed on: 20.10. 2015.

126. Tronco, V. M. 1997. **Manual for inspecting milk quality.** Santa Maria: UFSM Publishing House.

127. USDA - UNITED STATES DEPARTMENT OF AGRICULTURE. USDA **homepage**, 2010. Available at: <http://www.fas.usda.gov/psdonline>. Accessed on: 21.07.2013.

128. Valsechi, O. A. 2001. **Milk and its derivatives.** Technology of agricultural products of animal origin. Home agroindustry. Federal University of São Carlos Center for Agricultural Sciences Department of Agroindustrial Technology and Rural Socioeconomics. Araras. São Paulo. 17. Pp. Available at: http://www.cienciadoleite.com.br/?action. Accessed on: 20.04.2013.

129. Vargas, T.2003. **Quality and safety of milk and milk products**. III Foro Venezolano de La Leche. Available at: http://www.cavilac.org. Accessed on: 18.03.2013.

130. Viana, L. F. 2012. **Description of the flowchart and evaluation of some defects in mozzarella cheese.** Federal University of Goiás School of Veterinary and Animal Science Postgraduate Program in Animal Science. Seminar presented to the Postgraduate Program in Animal Science at the Veterinary and Zootechnical School of the Federal University of Goiás.

131. Vidor, M. V. S. 2003. Mini mobile milk processing plant: **evaluation of the effects on product quality and the reasons for family farmers joining.** Dissertation (Master's Degree in Agroecosystems) - Center for Agricultural Sciences - Federal University of Santa Catarina, Florianópolis.

132. Walstra, P.; Geurts, T. J.; Noomen, A.; Jelema, A. Van Boekel, M. A. J. S.; 1999. **Dairy technology: principles of milk properties and**

processes. Food science and technology. Marcel Dekker, Inc. New York - Basel. 727p.

133. Winck, C. A.; Scarton, L. M.; Capes, B.; Saggin, K. D. Machado, J. A. 2010. **Raw milk quality standards in Brazil**: international market insertion or social exclusion. University of Western Santa Catarina - UNOESC. CEPAN/UFRGS. Brazil.

134. Zumbado H. 2002. **Chemical analysis of foods**: Classical methods. Institute of Pharmacy and Food. Universidad de la Habana. p. 421-42

Annex A: Animal Production. Scientific Research Survey Form for the MSc in Food Production and Technology

PHASE 1: CHARACTERIZATION OF THE AGRI-LIVESTOCK PRODUCTION CHAINS OF ANGOLA

1- Location

Período ____

Província ____

2- List the animal species used in Angola for food production:

Fresh and fermented milk, yogurt and cheese

Municipality	Date	Species	Scientific Name	Varieties found	Common name	Plants and other places where they grow	Harvest time	Post-harvest processing	Fresh product sales and price	Processed products	Product sales and price	Waste produced	Destination

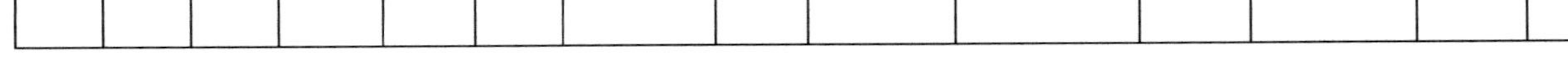

Raw milk powder   Raw milk powder

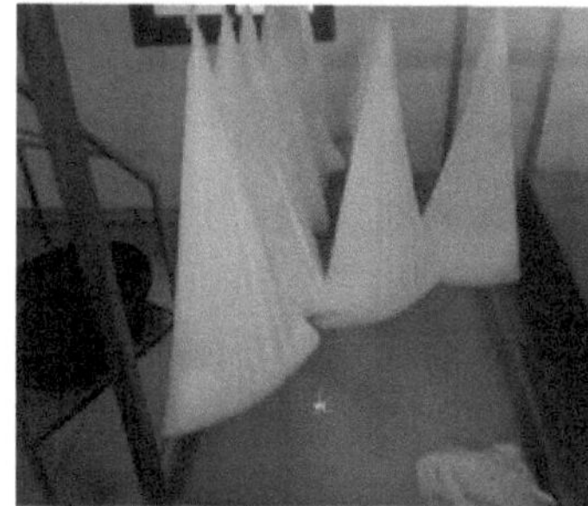

Desorption   Cheese ripening process AN

Cheese maturing room Cheese ready for sale

Liquid yogurt packaging Product storage room

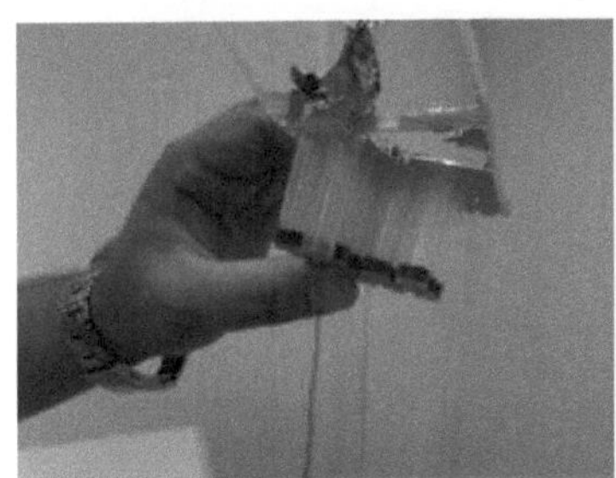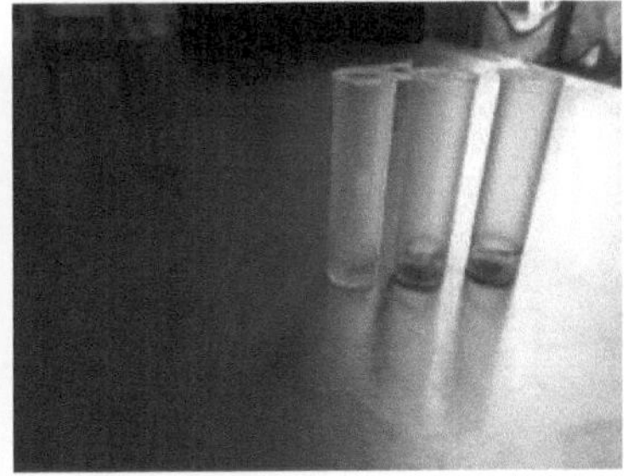

Test to determine the presence of antibiotics in milk

Raw material placed in the mill Mixing tank (sugar)

Plastic packaging machine

Liquid yogurt   "Omavele

Annex D - Physico-chemical analysis

Sample preparation and weighing

Determining acidity

Determination of pH

Determining the pH of butter Determining the pH of butter

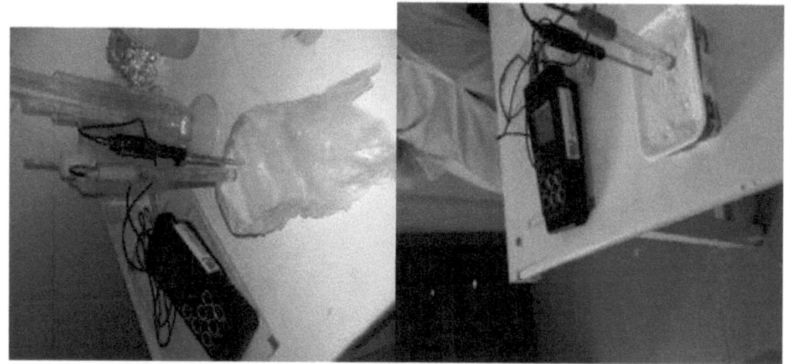

Determining the pH of cheese

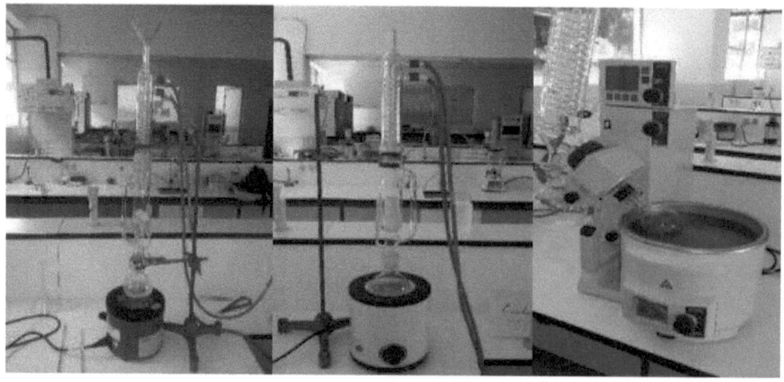

Fat determination by Soxhlet method

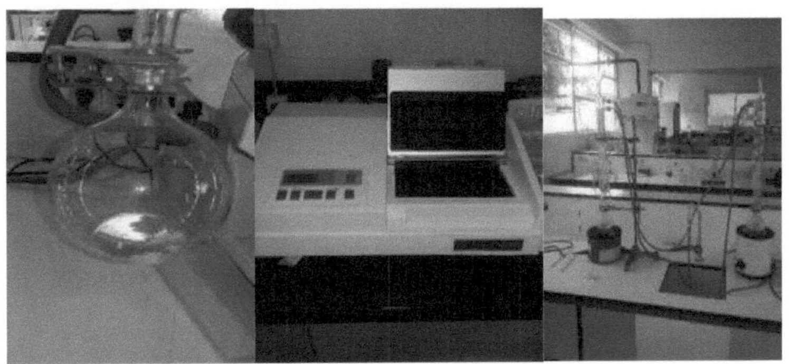

Protein determination

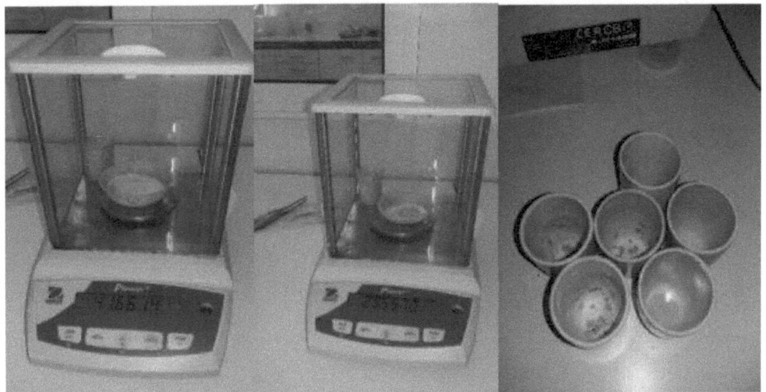

Ash determination

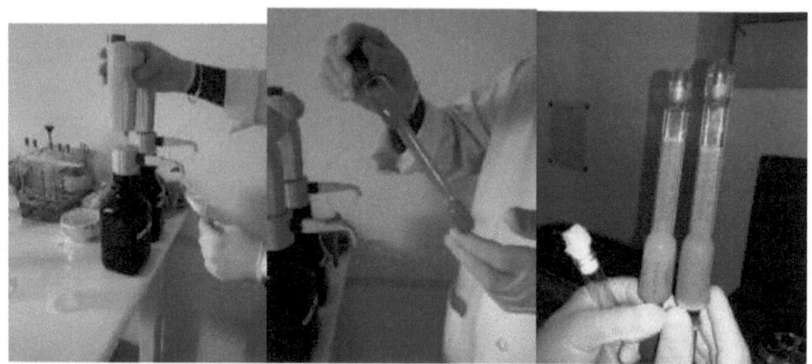

Determination of fat using the Gerber method

Printed by Books on Demand GmbH, Norderstedt / Germany